International Series in Operations Research & Management Science

Volume 300

More information about this series at http://www.springer.com/series/6161

Lotfi Azzabi • Dorra Azzabi • Abdessamad Kobi

The Multi-Criteria Approach for Decision Support

An Introduction with Practical Applications

Springer

Lotfi Azzabi
ESSCA School of Management
Angers, France

Dorra Azzabi
University of Angers
Angers, France

Abdessamad Kobi
Polytech Angers University
Angers, France

ISSN 0884-8289 ISSN 2214-7934 (electronic)
International Series in Operations Research & Management Science
ISBN 978-3-030-57261-7 ISBN 978-3-030-57262-4 (eBook)
https://doi.org/10.1007/978-3-030-57262-4

This Springer imprint is published by the registered company Springer Nature Switzerland AG
The registered company address is: Gewerbestrasse 11, 6330 Cham, Switzerland

Contents

List of Figures

List of Tables

Chapter 1
The Multi-Criteria Approach Decision

1.1 Introduction

Multiple-criteria decision-making or multiple-criteria decision analysis is a subdiscipline of *operations research* that explicitly considers multiple criteria in decision-making environments. Whether in our daily lives or in professional settings, there are typically multiple conflicting criteria that need to be evaluated in making decisions. Cost or price is usually one of the main criteria. Some measure of quality is typically another criterion that is in conflict with the cost. In purchasing a car, cost, comfort, safety, and fuel economy may be some of the main criteria we consider (Roy 1990). It is unusual to have the cheapest car to be the most comfortable and the safest. In *portfolio management*, we are interested in getting high returns but at the same time reducing our risks. Again, the stocks that have the potential of bringing high returns typically also carry high risks of losing money. In service industry, customer satisfaction and the cost of providing service are two conflicting criteria that would be useful to consider.

1.2 Decision Support

It is common in a study of decision support; have to consider several points of view to compare the relative attractiveness of different actions to solve the decision problem under consideration. The decision support using techniques and methodologies from the field of applied mathematics such as optimization, statistics, decision theory, and theories of less formal areas such as analysis of organizations and science cognitive.

L. Azzabi et al., *The Multi-Criteria Approach for Decision Support*, International Series in Operations Research & Management Science 300,
https://doi.org/10.1007/978-3-030-57262-4_1

1.2.1 Define Decision Support

Roy (1990) defines decision support as "the activity from that, building on models made explicit but not necessarily completely formalized, helps to get some answers to questions that arise one involved in a decision process, components used to inform the decision to prescribe and normally, or simply to promote a behavior that increase the coherence between the evolution of the process on the one hand, the objectives and value system the service which the worker is placed on the other." Thus defined, the decision support is not only very partial to the search for truth. Theories or, more simply, methodologies, concepts, models, techniques that support it have, in most cases, a different goal: preparing for the change as a decision-making process so as to increase its consistency with objectives and value system of one for whom or on whose behalf the decision aid is exercised. The theories or, more simply, the methodologies, concepts, models, techniques that support it have, in most cases, a different purpose: to prepare for change as a decision-making process in order to increase its consistency with the goals and value of the system. The one for whom or on behalf of whom the decision support is exercised. "A decision problem is not an object that already exists; the wording generally cannot be completely objective and cannot be considered independently of the relationship between the individual and reality" Roy (1998).

1.2.2 The Problem of Decision

The problem of decision may be seen as a focus of the investigation we adopt for a given decision problem. It expresses the terms in which the decision maker is the problem and reflects the kind of prescription they want. Description of the problem (P. δ): It is simply a question of describing the actions and their consequences, not to compare them as is the case with the other three problems above. Roy (1985).

- *Problem-choice (P. α):* It is to select a subset as small as possible for all actions A, containing the best actions. The ideal is to obtain a single best action. But because of the confrontational nature of the criteria, it is preferable to provide the decision maker a few shares that represent different variants of the "best action." Obviously, the end result can be refined using additional information or further analysis (E Costa and Vansnick 1994).
- *Problem-sorting (P.β)*: It is to assign each action to a set of predefined categories. This formulation is appropriate when the decision problem is to consider each action independently of the other (taking into account the intrinsic characteristics of each share) in order to propose a recommendation from a set of recommendations specified in advance. Each recommendation may be associated with a category. The decision problem is then seen as potential actions to sort the categories defined in terms of predefined standards. The sorting procedure should be defined so that each action is assigned to one and only one category. Formally, a prescription is a partition of A (Mousseau et al. 2000).

- *Storage Problem (P.γ)*: It is to store the various actions in action from the best to worst. This issue is interesting when the shares are differentiated by their relative. The ideal is to get a complete order. However, because of the confrontational nature of the criteria, the vagueness, the existence of different value systems, it is often more realistic to present the decision maker a partial order. It should be noted that in practice, the storage may be necessary only for the most interesting actions. Formally, the prescription is a partial order, a transitive relation defined on A (or a subset of A) (Mousseau et al. 2000).
- *Problem-description (P. δ)*: It is simply to describe the actions and their consequences, not to compare them as is the case with the other three issues above. Here, there is not a requirement and procedure of investigation is cognitive (Roy 1985).

1.2.3 Type of Criteria

A criterion is used to evaluate and compare potential actions in a well-defined goal. Each criterion is associated with a given unique descriptive. Several types of criteria:

- *The true criterion*: We talk about true test when the action that receives the highest score will be preferred to another. There is indifference only if both scores are equal (Fig. 1.1).
- *The quasi-criterion*: We talk about almost criterion when there is a range of indifference [−q, q]. There is indifferent between actions a and b where the difference of their assessment falls in the range [−q, q] (Fig. 1.2).
- *Pre-criterion*: A pre-test is a real test that introduces a low preference range. A multi-criteria decision method can be either:

The appearance of a compensatory method reflects the greater or lesser ability to offset a disadvantage in a test state that any method of multi-criteria decision can be either:

- *Compensatory*: in this case, we assume an absolute offset between the different evaluations. Thus, a good performance on a criterion can be easily offset by poor performance on another criterion. Several methods are in this category such as the weighted sum method.
- *No compensation*: no compensation is agreed between the different dimensions. The decision maker may state that the dimensions are large enough to refuse any kind of compensation or compromise. Include the lexicographic method.
- *Partially offsetting*: in this case, a kind of compensation is agreed between the different dimensions or criteria. Most multi-criteria methods are in this category. The major problem is to assess the degree of compensation of each method (Azzabi et al. 2009) (Fig. 1.3).

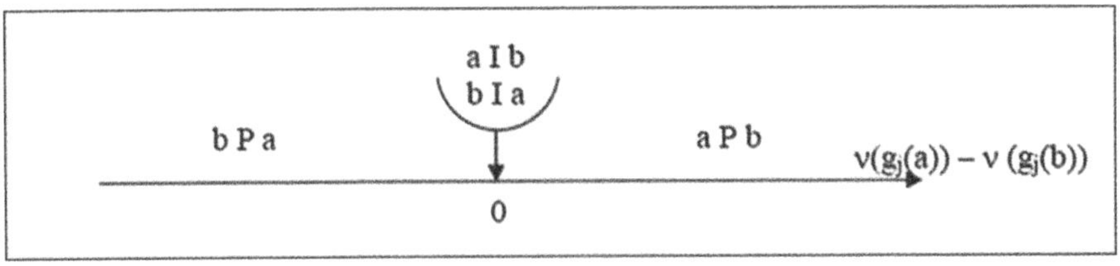

Fig. 1.1 The true criterion

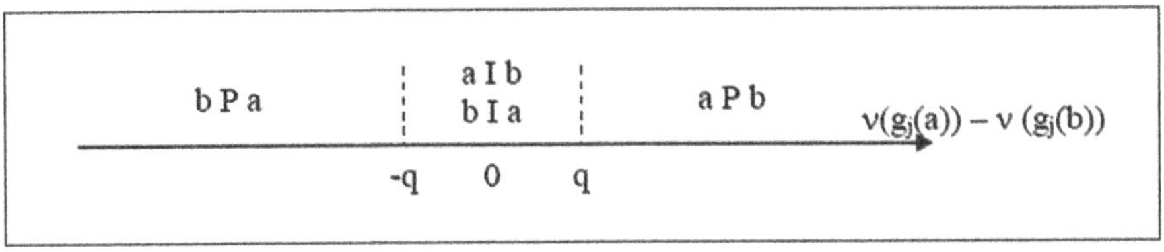

Fig. 1.2 The quasi criterion

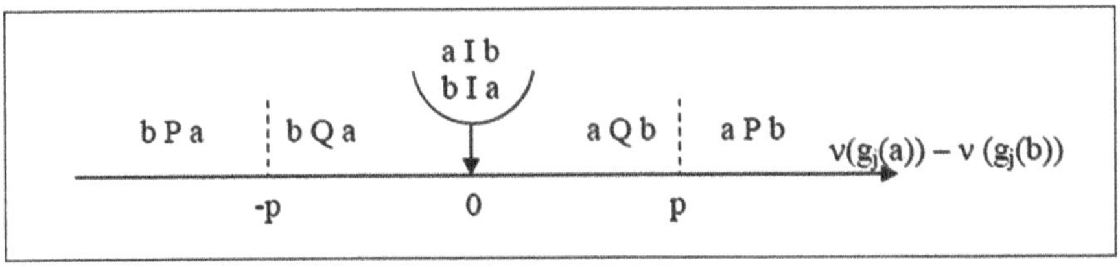

Fig. 1.3 The pre-criterion

1.3 What Is MCDA (Multi-Criteria Decision Analysis)?

A form of MCA that has found many applications in both public and private sector organizations is multi-criteria decision analysis or MCDA for short (also known as multi-attribute decision analysis, or MADA). This chapter explains what MCDA is and then outlines what is required to carry out such an analysis.

MCDA is both an approach and a set of techniques, with the goal of providing an overall ordering of options, from the most preferred to the least preferred option. The options may differ in the extent to which they achieve several objectives, and no one option will be obviously best in achieving all objectives. In addition, some conflict or trade-off is usually evident among the objectives; options that are more beneficial are also usually more costly, for example. Costs and benefits typically conflict, but so can short-term benefits compared to long-term ones, and risks may be greater for the otherwise more beneficial options.

The technical aspects of MCDA is a way of looking at complex problems that are characterized by a mixture of monetary and non-monetary goals, breaking the problem down into more manageable pieces. The purpose is to serve as a decision-making aid, but not to make the decision. As a set of techniques, MCDA offers different ways to simplify a complex problem. The purpose is to serve as an aid to thinking and decision-making, but not to take the decision. As a set of techniques, MCDA provides different ways of disaggregating a complex problem, of measuring the extent to which options achieve objectives, of weighting the objectives, and of reassembling the pieces.

The first complete exposition of MCDA was given in 1976 by Keeney and Raiffa. They built on decision theory, which for most people is associated with decision trees, modeling of uncertainty, and the expected utility rule. By extending decision theory to accommodate multi-attributed consequences, Keeney and Raiffa provided a theoretically sound integration of the uncertainty associated with future consequences and the multiple objectives those consequences realize.

The main assumption embodied in decision theory is that decision makers wish to be coherent in taking decisions. That is, decision makers would not deliberately set out to take decisions that contradict each other. No one would place several bets on the outcome of a single race such that no matter which horse won they were certain to lose money. The theory expands on this notion of coherence, or consistency of preference, and proposes some simple principles of coherent preference, such as the principle of transitivity: if A is preferred to B, and B to C, then A should be preferred to C, which is a requirement if preference is to be expressed numerically. By treating these rather obvious principles as axioms it is possible to prove non-obvious theorems that are useful guides to decision-making. A parallel can be found in the study of geometry. Simple principles like "The shortest distance between two points is a straight line" are combined using the rules of logic to prove theorems that are not obvious, like the Pythagorean principle, that the square of the hypotenuse equals the sum of the squares of the other two sides.

The first two theorems establish a logical equivalence between coherent preference and number systems. If preferences are coherent, then two sorts of measures follow logically: probability and utility both associated with the consequences of decisions. The first theorem establishes the existence of probabilities: numbers which capture the likelihood that consequences will occur. The second theorem shows the existence of utilities: numbers which express the subjective value of the consequence and the decision maker's risk attitude.

The third theorem provides a guide to taking decisions: choose the course of action associated with the greatest sum of probability-weighted utilities.

That is the expected utility rule, which has existed in various guises for over 200 years. To apply the expected utility rule, assess a probability and utility for each possible consequence of a course of action, multiply those two numbers together for each consequence, and add those products to give the expected utility for that course of action. Repeat the process for each course of action, and choose the action associated with the largest expected utility.

That description sounds rather dry and impractical, but decision theory gave birth to the applied discipline of decision analysis (Howard 1966).

The use of MCDA by various governmental agencies in the United States, at local, state, and federal level, is now widespread. The approach has also withstood challenges of its results in courts of law and inquiries. The audit trail left by a well-conducted MCDA suits the climate of freedom of information in the United States, though it has not always been favored by those who wished to take very different decisions from those recommended by the analysis. A simple example is: Analysis of alternative sites for the storage of nuclear waste in the United States. Five potential sites were analyzed using MCDA, resulting in an overall ranking of sites. The subsequent announcement by the US Department of the Environment of three sites for further investigation included the sites ranked first, third and fifth by the MCDA rather than the top three (Keeney and Merkhofer 1987).

1.4 The Stages of a Multi-Criteria Analysis

Multi-criteria analysis is described here as a cut and dried, step-by-step process. However, unless the user has applied the method to very similar problems in the past, it is more appropriate to envisage it as a guided exploration of a problem. Some of the steps will require detailed thought about issues surrounding the decision.

Much of the value derives from the thought that goes into the early steps (Fig. 1.4).

1.4.1 Establishing the Decision Context

A first step is always to establish a shared understanding of the decision context. The decision context is the whole panoply of administrative, political, and social structures that surround the decision being made. Central to it are the objectives of the decision-making body, the administrative and historical context, the set of people who may be affected by the decision, and an identification of those responsible for the decision.

It is crucial to have a clear understanding of objectives. To what overall ambition is this decision seeking to contribute? MCA is all about multiple conflicting objectives. There are ultimately trade-offs to be made.

Nonetheless, in applying MCA it is important to identify a single high level objective, for which there will usually be sub-objectives.

To establish objectives (and criteria) we need to establish both who the decision makers are (in order to establish objectives) and also people who may be affected by the decision.

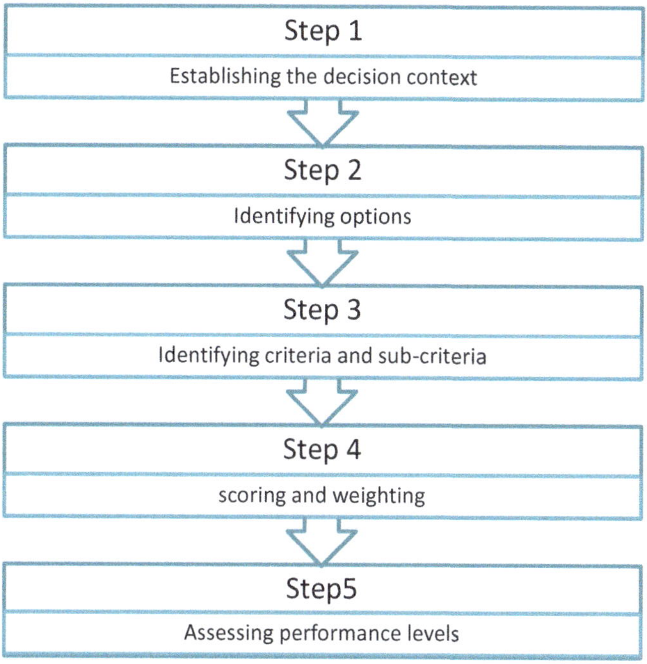

Fig. 1.4 Steps in a multi-criteria analysis

1.4.2 Identifying Options

Having established the decision context, the next step is to list the set of options to be considered.

It is unlikely, even with a new and unexpected problem, that the decision-making group will arrive at the stage of formal MCA structuring without some intuition about options. Often in practice there will be ideas, sometimes going back many years. Sometimes the problem will be an embarrassment of possibilities and it will be the role of the MCA in the first instance to provide a structured sifting of alternatives to identify a short-list, using basic data and quick procedures.

It is sometimes worth carrying out some informal sifting against established legal and similar restrictions. It is not worth considering and putting effort into gathering data about clearly infeasible propositions.

The first visit to step 2 may well not be the last, particularly in problems where there is a paucity of acceptable alternatives. The later steps of the MCA may demonstrate that none of the alternatives is acceptable and can serve to crystallize thoughts as to where the inadequacies lie. At this stage, fresh ideas and creative thought are needed. This will be informed by the MCA. For example, it may encourage a search for new options that combine the strong points of one existing option in some areas with the strong points of another in a different area.

The failure to be explicit about objectives, to evaluate options without considering what is to be achieved; options are important only for the value they create by achieving objectives. It might be better to consider objectives first, particularly when the options are not given and have to be developed.

1.4.3 Identifying Criteria and Sub-Criteria

1.4.3.1 Overall Approach

The criteria and sub-criteria are the measures of performance by which the options will be judged. A large proportion of the "value-added" by a formal MCA process derives from establishing a soundly based set of criteria against which to judge the options.

Because the criteria serve as the performance measures for the MCA, they need to be operational. A measurement or a judgment needs to specify how well each option meets the objectives expressed by the criteria. We shall return to this later, but a question to be borne in mind in developing the set of criteria is "Is it possible in practice to measure or judge how well an option performs on these criteria?"

1.4.3.2 Procedures to Derive Criteria

Whether in a decision-making team or as an individual, an effective way to start the process of identifying criteria is first briefly to recapitulate step 1 and then to brainstorm responses to the question "What would distinguish between a good choice and a bad one in this decision problem?"

Responses should all be noted down uncritically, perhaps on whiteboards if in a group context.

Interest group perspective(s) may be important. One way to include them is directly to involve the affected parties in some or all stages of the MCA. This might be appropriate, for example, in some local planning issues. A second approach is to examine policy statements and secondary information sources from the various interest groups and to analyze these to derive criteria to reflect their concerns. A third approach, if the decision-making team has experience in decision-making, it helps to encourage one or more of its members to act as key interest groups, to ensure that this perspective should not be overlooked when developing the criteria.

Often, both decision-maker objectives and interest group viewpoints may be articulated in broad-brush terms. For example, a criterion like environmental impact might be suggested. In many circumstances, assessing options against such a broad criterion may prove difficult, even though the notion of environmental impact may be important.

Typically, in the process of eliciting criteria, after an initial hesitation, suggestions come thick and fast, until eventually the process slows and dries up. At the end of a relatively short period, it is normal to have a substantial list of potential criteria.

The number of criteria should be kept as low as is consistent with making a well-founded decision. There is no "rule" to guide this judgment and it will certainly vary from application to application. Large, financially or otherwise important choices with complex technical features (such as a decision on where to locate a nuclear waste facility) may well have upwards of a hundred criteria. More typical, however, is a range from six to twenty.

1.4.3.3 Grouping Criteria

It can be helpful to group together criteria into a series of sets that relate to separate and distinguishable components of the overall objective for the decision. This is particularly helpful if the emerging decision structure contains a relatively large number of criteria (say eight or more).

The main reasons for grouping criteria are: (a) to help the process of checking whether the set of criteria selected is appropriate to the problem; (b) to ease the process of calculating criteria weights in large MCDA applications, when it can sometimes be helpful to assess weights firstly within groups of related criteria, and then between groups of criteria; and (c) to facilitate the emergence of higher level views of the issues, particularly how the options realize trade-offs between key objectives.

For both these reasons, grouping criteria is an important part of an MCA. However, there are few formal guidelines to determine what a "good" structure is and what is "bad." Most experienced decision analysts see problem structuring as a skill that is acquired primarily through practical experience. For most large problems, there is arguably no unambiguously correct structure or grouping of criteria.

An acceptable structure is simply one that reflects a clear, logical, and shared point of view about how the many criteria that may be relevant to an MCA assessment can be brought together into coherent groups, each of which addresses a single component of the overall problem. For example, in assessing forms of medical intervention for a given condition, one group of criteria may relate to the patient's experience. Often the criteria in an MCA reflect individual measurable indicators of performance relative to the issue at hand, whereas the groups of criteria reflect sub-objectives to the single main objective that underlies the MCA process.

While knowledge of the domain of the particular problem will often give very clear guidance as to what are clear and helpful groups of criteria, there can be room for debate. For example, should criteria relating to the time to the end of the treatment (speed of admission, length of stay) constitute one sub-objective, with criteria reflecting the experience of the treatment itself being placed in their own cluster? To some extent, such debate is helpful and to be expected. It is one way in which the decision makers explore the problem to be solved and come to a shared understanding of its characteristics and what factors should drive their choice.

Especially in some complex and contentious public sector decisions, it is likely that different stakeholder groups, because of their very different ways of framing the problem, may have substantial difficulty in sharing the same grouping of criteria. It may need prolonged negotiation and a good deal of tact and ingenuity to arrive at a shared structure (Brownlow and Watson 1987).

1.4.3.4 Assessment of the Provisional Set of Criteria

Before finalizing the choice of criteria the provisional set needs to be assessed against a range of qualities.

Completeness

Have all important criteria been included? This needs some care as it is not necessarily obvious from the beginning what the important criteria are of the MCA?

Redundancy

Are there criteria which are unnecessary? In principle, criteria that have been judged relatively unimportant or to be duplicates should have been removed at a very early stage, but it is good practice to check again. The MCA team may also wish to delete a criterion if it seems that all the available options are likely to achieve the same level of performance when assessed against it. If this were the case, then omitting it would not affect any ranking of options and would economize on analytical input. However, omission on these grounds should be approached with care. First, it has not yet been formally assessed how well each option will show up on the criterion concerned. Secondly, it may be that at a later stage new options come into consideration which do not exhibit this behavior, especially in MCA systems that may be used by delegated groups and/or to address a series of different problems.

Operationality

It is important that each option can be judged against each criterion. The assessment may be objective, with respect to some commonly shared and understood scale of measurement, like weight or distance. Optionally, it can be judgmental, reflecting the subjective assessment of an expert. A strength of MCA is its ability to accommodate and use simultaneously both forms of assessment of options. In either case, however, the criterion must be defined clearly enough to be assessed. It can sometimes be helpful to break a criterion down into a further sub-level of more explicitly defined criteria, if assessment at a particular level is problematic.

Mutual Independence of Preferences

Straightforward applications of MCA require that preferences associated with the consequences of the options are independent of each other from one criterion to the next. The key idea is simple: can you assign preference scores for the options on one criterion without knowing what the options' preference scores are on any other criteria? If the answer is yes, then this criterion is preference independent of the others. The question then is asked for each of the remaining criteria in turn. If the answer is always yes, then the criteria are considered to be mutually preference independent. This condition has to be met if the sum of weighted averages is to be used to combine preference scores across criteria, and this is true for all MCA approaches, whether they recognize it formally or not. Preferences are not always mutually independent. For example, the enjoyment a person gets from consuming a trifle may not be the sum of the amount of jelly, custard, sponge, etc., it contains, but be related in some way to the proportions in which they are combined. If this is the case, a simple weighted sum of the amounts of jelly, custard, and so forth contained in a set of option trifles will not in general reproduce the preference ranking that the individual has for the trifles.

In practical terms, the preferential independence question may be approached by asking, for each criterion, whether the preference scores of an option on one criterion can be assigned independently of knowledge of the preference scores on all the other criteria.

If the answer is no, then it may be necessary to use more complex models for combining scores across criteria. However, two simpler approaches may be possible. The first is to combine the two criteria that are not preference independent of each other into a single criterion, which captures the common dimension of value. This will be effective provided that the new criterion is itself preference independent of the remaining criteria. The second approach is to recognize that options often have to satisfy a minimum acceptable level of performance for them to be considered; options falling below any minimum level are rejected outright because better performance on other criteria cannot compensate. This hurdle usually guarantees preference independence of the criteria; all options fall at or above the minimum level of performance, so that preference on any given criterion is unaffected by preference on the others. If preference independence is still violated, then more advanced MCA procedures must be adopted.

Double Counting

Public sector decisions can be particularly prone to double counting, especially of effectiveness or benefits. This stems, for example, from a desire to set out the distribution of effects on different parts of the population. As a consequence, it is quite easy for the same basic impact to be recorded more than once in a performance matrix. As with CBA, double counting should not be allowed in MCA, since double-counted effects are likely to be given more weight in the final overall decision than they deserve.

On occasions, however, it is desirable to present the same basic effect from more than one point of view, so that the overall context of the decision is fully understood by those involved. For example, both numbers of accidents saved and the money value of accident cost savings are sometimes recorded as separate items in appraisal of transport schemes. It is important, however, when moving from this multi-perspective form of presentation of options on to the process of choice between options, that the potential for any double counting is recognized and the final performance matrix used for decision-making is suitably amended to remove it.

Sometimes, what appears to be double counting is not.

A good test is to ask the following question: "Two compounds, A and B, will cost the same to develop, are expected to yield the same financial return, and are identical on all other criteria except that A meets a greater unmet medical need than B. Do you prefer A or B, or are you indifferent?" If the answer is a preference for A, then there must be more to the value associated with A than its expected commercial value. Exploring the reasons for the preference will uncover additional criteria.

Size

An excessive number of criteria lead to extra analytical effort in assessing input data and can make communication of the analysis more difficult. A final check to ensure that the structure is no larger than it needs to be is useful at this stage.

In a full application of MCA, criteria are explicitly weighted. However, in the absence of weighting, if there is any possibility of informal judgments being made by scanning the performance matrix, it is wise at this stage to ensure that marked inconsistencies between the number of criteria and the likely importance of the topics they reflect are, if possible, eliminated. If this is not practicable, then particular care needs to be taken to prevent the imbalance distorting people's interpretation of the matrix.

Impacts Occurring Over Time

Many public sector decisions concern expenditures to be undertaken now that will have impacts occurring over several subsequent years. Although this can present some difficulties in aggregating all effects into a single measure, with monetary-based techniques discounting is a reasonably well established procedure for aggregation. In MCA there is no single equivalent technique, although in principle the conventional discounting of money values can be accommodated and it can also be applied to physical impact indices other than monetary value. The reasons why these are not generally done are probably more cultural than substantive. Certainly, good decision facilitating practice would ensure that participants in any decision-making exercise had their attention drawn to time-differentiated impacts and gave thought to how these were to be consistently accommodated in the assessment. If a target completion date is an important consideration, it can be modeled as a separate criterion, with a non-linear value function. Options that are expected to deliver on time get good scores, those expected to deliver slightly late receive lower scores, and very late ones get zeros. Time has to be included in the definition of many other criteria so that temporary consequences can be distinguished from permanent ones. This is usually done by being explicit about the time horizon over which the consequences are being valued. Time horizons may differ from one criterion to the next, e.g. separately identifying short-term and long-term health effects. A further possibility would be to use some other principle of giving less importance to impacts in the long-run future. Alternatively there are approaches supported by some environmentalists for giving greater influence to longer term impacts. Finally, it would be possible to carry out an MCA.

1.4.4 Scoring and Weighting

1.4.4.1 The Performance Matrix Without Scoring and Weighting

A basic MCA will present the decision maker with the performance matrix itself. The task for the decision maker is then to study the matrix, and come to a view on the ranking of the options—probably assisted by some supplementary advice from those who constructed the matrix on their views of how the information should be interpreted.

The measures used in performance matrices of this kind are often qualitative descriptions (for example, of styling), or natural units (such as price or length), or sometimes a crude numerical scale (e.g. number of stars), or even a scale of 0 to 100.

For government applications the use of 0 to 100 numerical scales is not recommended if the analysis is not to proceed to the numerical analysis. The extra work entailed in producing such scales can all too easily be counterproductive, by giving the intuitive but incorrect message that the scores can then be added together.

Even if the matrix is confined to qualitative description, natural units, and very simple scales (such as stars), it is advisable to try to use similar numbers of criteria within each major sector of the value tree.

1.4.4.2 Judgements Between Options Without Scoring and Weighting

To what extent does a performance matrix alone allow any comparison of options? The basic answer is only a little. What is perhaps just as important is to be clear about what types of comparison may not be made, and why.

Dominance

First, it is possible to examine the set of options to check for the presence of dominance.

Assuming that all the estimates of criteria scores are accurate, if option A dominates option B, then B cannot be the single best one available. Thus, if the purpose of the MCA is to recommend a single best option, B may be removed from consideration. If the purpose is short-listing, then it is possible, but rather unlikely, that a dominated option would be taken through to a later stage in the selection process. Logically it would only make sense to do so if it was thought that new information about options might come to light, that some of the criteria scores might be inaccurate, or if there was some possibility that the dominating option (A) might cease to be available.

For screening, being dominated might exclude an option from further consideration, depending on the number of options required for later consideration and the strength of the others available, but dominance says only that B must rank lower than A.

Finally, it can be noted that dominance is transitive. If A dominates B, and B dominates C, then A will always dominate C and this dominance does not need to be directly checked.

In practice, dominance is rare. The extent to which it can help to discriminate between options and so to support real decisions is correspondingly limited.

Other Approaches

Dominance is limited in the extent to which it can differentiate between options specifically because it makes no assumption at all about the relative importance of criteria (the different columns), nor does it employ any supplementary information beyond that directly displayed in the performance matrix. It is a common (although often sub-conscious) intuitive misuse of the performance matrix to either:

(a) Add recorded performance levels across the rows (options) to make some holistic judgment between options about which ones are better;
(b) Eliminate (or prioritize) options that record weak (or strong) performance levels on particular criteria.

In the first case, the implication is that all criteria contribute with equal importance to options' overall performance, when this has not been established. In (b) the same error is potentially made (in deciding the criteria where strong or weak performances are identified) and there may additionally be a degree of arbitrariness in deciding the thresholds for recognizing a special level of performance. In addition to dominance, there are a limited number of non-compensatory, MCA procedures that may be applied, using supplementary information of various sorts beyond that in the performance matrix.

The Limitations of Human Judgments

Research on human judgments and decision-making (Bazerman 1998) shows that the simplifications which we make to enable us to deal with complex problems sometimes do not work well. We are inclined, for example, to be biased in our assessments of alternatives that can more readily be linked to what is familiar (the "representativeness heuristic"), and to be unduly influenced by recent, memorable, or successful experience (the "availability heuristic").

MCA techniques are designed to help overcome the limitations by imposing a disciplined structure which directs attention to criteria in proportion to the weight which they deserve.

The development of a performance matrix is an important step in this direction, but it is limited because a subjective interpretation of the matrix is still prone to many of these well documented distortions of human judgments, as well as the intuitive processing errors. (Communities and Local Government 2009)

In practice, the extent to which options can be compared using non-compensatory methods is strictly limited. The alternatives at this stage are either to end the MCA, reverting to an informal treatment of the decision, for which the performance matrix simply provides basic factual information in a more considered and coherent way than might otherwise have been the case, or to move on to a formal, compensatory MCA.

1.4.5 Assessing Performance Levels

The first consideration in setting up consistent numerical scales for the assessment of criteria is to ensure that the sense of direction is the same in all cases, so that (usually) better levels of performance lead to higher value scores. This may mean a reversal of the natural units.

It is conventional to allot a value score to each criterion between 0 and 100 on an interval scale. The advantage of an interval scale is that differences in scores have consistency within each criterion, although it does not allow a conclusion that a score of 80 represents a performance which on any absolute standard is five times as good as a score of 16 (which would require a ratio scale of measurement). The "ruler" which the scoring scale represents is good only within the confines of this particular MCA. However, when combined with appropriately derived importance weights for the criteria, the use of an interval scale measurement does permit a full MCA to be pursued.

The first step in establishing an interval scale for a criterion is to define the levels of performance corresponding to any two reference points on the scale, and usually the extreme scores of 0 and 100 would be used. One possibility (global scaling) is to assign a score of 0 to represent the worst level of performance that is likely to encounter in a decision problem of the general type currently being addressed, and 100 to represent the best level.

Another option (local scaling) associates 0 with the performance level of the option in the currently considered set of options which performs least well and 100 with that which performs best.

The choice between local and global should make no difference to the ranking of options. An advantage of global scaling is that it more easily accommodates new options at a later stage if these record performances that lie outside those of the original set. However it has the disadvantages of requiring extra, not necessarily helpful judgments in defining the extremes of the scale and, as will be seen in the next chapter, it lends itself less easily than local scaling to the construction of relative weights for the different criteria. (Communities and Local Government 2009)

Once the end points are established for each criterion, there are three ways in which scores may be established for the options.

1.5 Different Types of MCA

An important initial consideration in the choice of MCA technique is that of the number of alternatives to be appraised. Some problems, especially in design and engineering, are concerned with outcomes that are infinitely variable. However, most policy decisions, even at fairly low levels, are usually about choices between discrete options, for example, between alternative investment projects, or between alternative types of tax system.

Where the number of options is finite, it does not matter in principle whether this number is small or large. However, it is important to bear in mind that each option that has to be considered has to be appraised to determine how well it performs on each of its criteria. Gathering and processing these data will consume resources, the more so if a large number of criteria have been identified. In choosing whether to implement one of the simpler or one of the more detailed MCA decision support procedures, this is a factor to bear in mind.

The multi-criteria analysis methods or, more accurately, methods of multi-criteria decision are techniques that allow you to integrate any type of criteria; these procedures seem more possible to move towards a judicious compromise rather than optimum often obsolete (Communities and Local Government 2009).

1.5.1 Non-compensatory Methods

This method try to establish preferences between options when:

- Each option is evaluated against a common set of criteria set out in a performance matrix.
- The decision maker is not willing to allow compensation, for strong performance on one criterion to compensate for weak performance on some other criterion.

Commitment to non-compensatory evaluation severely restricts the extent to which, in practice, overall preferences between options can be established.

1.5.1.1 Conjunctive and Disjunctive Selection Procedures

If the decision maker is prepared to allow one or more supplementary judgments to complement the information in the performance matrix, then it is possible to go a little further with non-compensatory decision-making about options.

One possibility is to allow the introduction of thresholds of performance for one or more criteria. The extra information introduced consists of the thresholds themselves and, implicitly, a judgment that the criteria concerned justify being prioritized in this way over others for which no thresholds may be given. What is affected is a form of external benchmarking.

The conjunctive model eliminates options that fail to reach externally set levels of performance on each of one or more named criteria.

The disjunctive model allows an option to pass if the option concerned meets a minimum threshold level of performance on at least one of a set of named criteria. Continuing the selection example, this would be equivalent to allowing candidates through to the short-list provided they exhibited potential in terms of any one of the key criteria for which thresholds have been set.

It is perfectly permissible to use both a conjunctive and a disjunctive filter in any one decision-making process, with different threshold levels, perhaps applied to different sets of criteria.

If the external benchmarks are set independently and represent accepted reasons, whereby an option that fails to "pass" would not normally ever be considered for implementation, then this type of screening process is valid, although it is arguable that it scarcely needs formal MCA to make the judgments. If, on the other hand, the performance thresholds are being set in the light of knowledge of what all the options are scoring in this particular analysis, then the process may induce unforeseen bias. By selecting particular criteria to be used for benchmarking (or by identifying benchmark levels themselves) an implicit judgment is being made about the relative importance of criteria, the basis for which may not have been fully thought through. (Communities and Local Government 2009)

General practical advice would be that disjunctive and conjunctive procedures may help in providing some structure and an audit trail in developing a long short-list from an initial list of candidate options. However, for later stages in the decision-making process, where candidate options are harder to separate in terms of overall performance, the MCDA model is likely to prove a more reliable guide.

1.5.1.2 Lexicographic Ordering

Another approach to non-compensatory choice requires the decision maker not to supply external benchmarks, but instead to provide supplementary information about the ranking of criteria in terms of perceived importance. It then considers each criterion in turn and works as a sequential elimination model.

Specifically, in lexicographic elimination, all options are first compared in terms of the criterion deemed most important. If there is a unique best performing option in terms of this criterion, then that option is selected as the most preferred. If there is a tie, then the selection process moves on to the second ranked criterion and, just for those options which tied for first previously, seeks the remaining option which scores best on the second criterion. Again, if this leads to a unique selection, then this option is designated the preferred one. If not, the process is repeated for options tying in terms of both the first and second ranked criteria, using the third criterion and so on until a unique option is identified or all criteria have been considered.

1.5.1.3 Elimination by Aspects

This model combines elements of both lexicographic ordering and the conjunctive/disjunctive models. Options are compared against a threshold. They are examined criterion by criterion and, for each criterion, options which do not pass the threshold are eliminated. However, criteria are examined not in order of importance, but in perceived order of likelihood of maximizing the number of options that fail to pass. This process is continued until only one option remains.

Neither lexicographic elimination nor elimination by aspects has contributed much to the practice of public sector decision-making.

1.5.2 Multi-Attribute Utility Models

Research into decision-making is divided into three principal streams:

- Descriptive: which examines how individuals, groups, and organizations actually undertake decision-making in practice?
- Normative: which tries to establish how rational individuals should choose between competing options, and
- Prescriptive: which, recognizing some of the weaknesses in intuitive, unaided human decision-making that descriptive decision research has identified, seeks procedures to bring decision-making in practice closer to normative ideals.

Although there are some clear links back to earlier work, mostly on the analysis of economic behavior and decision-making, the starting point for MCA in terms of a normative theory of how individuals should rationally choose between competing options is generally seen as the work of von Neumann and Morgenstern, (Von Neumann and Morgenstern 1947).

These authors aimed to derive a theory of how rational individuals ought to choose between options. They first established a set of fundamental axioms of rational choice. Then, by using mathematical reasoning, they showed that the only way an individual could behave consistently with the full set of axioms is by choosing the option which possessed the maximum subjective expected utility (SEU) value.

It is assumed that there is a range of separate options to choose between, that one and only one of these options must be chosen now, and that, because of uncertainty about exactly what the future will be, different options will have potentially different values (utilities) to the decision maker, depending on what kind of future (termed state of nature, or state of the world) eventually transpires. For example, the utility of investing in a new reservoir may depend on future rates of climate change.

The SEU of any option is obtained by

1. Identifying all future states of the world that could reasonably be viewed as relevant to this decision,
2. Calculating the utility (degree of attractiveness), u_{ij}, the decision maker associates with the outcome that follows from the combination of choosing option i (say) and it later turning out that future state of the world j actually occurs,
3. Creating the probability-weighted average of all the outcome utilities, where the probabilities are the individual's subjective estimates of the probability of each of the outcomes actually occurring.

$$u_i = p_1 u_{i1} + p_2 u_{i2} + \ldots + p_n u_{in} = \sum_{j=1}^{n} p_j u_{ij}$$

where

u_i: is the overall utility (preference score) of option$i$;

u_{ij}: is the utility of option i if, having chosen option i, it subsequently transpires that state of the world j occurs;

p_j: is the decision maker's best judgment of the probability that future state of the world j will occur. (Communities and Local Government 2009)

This basic model of rational choice under uncertainty, although not without its critics, is the single most broadly accepted normative model of rational choice. While it deals explicitly with uncertainty, it is also implicitly a multi-criteria model since each of the individual u_{ij} is in principle based on a multi-criteria assessment.

The SEU model, however, does not provide direct prescriptive decision support for the multi-criteria problems addressed in this manual. This is because it does not indicate how they u_{ij} should be evaluated. However, the SEU model does not provide direct normative decision support for the multicriteria problems discussed in this manual. Indeed, it does not indicate how to evaluate them, research on this subject did not take place before the publication of Keeney and Raiffa who presented a set of operational calculation procedures.

It is beyond the scope of this manual to go into full detail on procedures for calculating u_{ij} estimates, which can be relatively complex. For this, see, e.g., Keeney and Raiffa (1976) or Goodwin and Wright (1988) for an introductory treatment.

The extra complexity stems from the fact that the multi-attribute utility model is simultaneously seeking to take account both of uncertainty and evaluating in terms of several criteria.

1.5.3 Linear Additive Models

The linear additive multi-criteria model has a straightforward intuitive appeal and transparency that ensures it a central role in any discussion of MCA. In part for this reason, it provides the basis for the MCA.

However, as with many tools, it can be misused. In particular, this is critical with regard to the scaling of options' performances on criteria, the weighting of criteria, and the relationship between weight determination and the scales on which performance on each criterion is measured. Failure to follow the proper logic of the model can lead to an MCDA that appears clear and well-founded, but which is, in fact, misleading and not a true reflection of the decision-making group's understanding of the problem.

Although the MCDA model is not difficult to apply, for some time researchers have sought ways of using linear additive choice models that require less precise data inputs than the basic model. Of course, with less precise information input, the recommendations output by the model will be less precise too and less likely unambiguously to identify a "best" option.

However, it may be that this is a price worth paying, if the input demands on the decision maker are less, or if the model can be constructed and applied more quickly. (Communities and Local Government 2009)

The most common way to seek to accommodate less precise inputs is to allow specific weights or scores to be replaced by statements that put bounds on the values the inputs can take, and/or express restrictions on the relative magnitude of some of the input values. These restrictions are most often in the form of linear inequalities.

Early thinking along these lines was developed by Fishburn (1965) in the context of single criterion decision-making under uncertainty (circumstances mathematically analogous to the linear additive MCA model).

In the linear additive MCA model, there are two inputs, weights and scores. In principle, decision makers may well be uncertain about the accuracy of either w_j or s_{ij}, or both. In many cases but not all, it is the weights where confidence about accuracy is lower. This is because many of the criteria will be assessed on objective scales of measurement, or against a shared background of evidence and experience that will tend to focus decision makers towards broadly similar assessments.

The extent to which this process will finalize the recommendation of a single option will depend on how restrictive are the set of inequality restrictions on w_j and on the relative performance of the options. Often it will not be sufficient to determine a unique recommendation. Dominance in weighted score differences may eliminate some options. Thereafter a more judgmental process may be needed.

A second approach is to seek to identify a single set of weights that is representative of all the possible weight combinations that are admissible, consistent with the established linear inequality constraints on the weights.

The above assumes that decision-makers' weight estimates are imprecise, but that estimates of options' scores on each criterion are accurate. If the reverse is the case, then the first of the exploration and short-listing process described above (based on minimum and maximum weighted s scores and dominance in weighted scores) may be worth pursuing. In this case, however, it will generally be necessary to run a simple linear programming computer package to identify the required information about options' performance at the vertex points (Communities and Local Government 2009).

As an alternative, insights about variability in the overall performance levels of different options can also be obtained by applying simple risk analysis methods, where imprecision in the true values of individual performance scoress_{ij} can be represented by probability distributions.

1.5.4 Outranking Methods

The principal outranking methods assume data availability broadly similar to that required for the MCDA model. That is, they require options to be specified, their performance to be assessed on a series of criteria, and for weights to be assessed that express the relative importance of the criteria.

Outranking is a concept originally due to Roy (1974) and may be defined as follows.

Option A outranks Option B if, given what is understood of the decision-maker's preferences, the quality of the evaluation of the options and the context of the problem, there are enough arguments to decide that A is at least as good as B, while there is no overwhelming reason to refute that statement.

Thus outranking is defined fundamentally at the level of pair-wise comparison between every pair of options being considered.

Based on this rather general idea, a series of procedures have been developed to operationalize outranking as a way of supporting multi-criteria decision-making. Typically, they involve two phases. First, a precise way of determining whether one option outranks another must be specified.

Secondly, it is necessary to determine how all the pair-wise outranking assessments can be combined to suggest an overall preference ranking among the options.

The thinking behind outranking methods is quite different from that underlying MCDA. Although they aim to achieve broadly the same outcome, the two are not easy to compare in a way that does justice to both approaches. In essence, the MCDA approach makes relatively strong assumptions about the underlying circumstances of the problem and delivers in a quite formulaic way a ranking of options, which may then need to be considered carefully to avoid overlooking important factors not captured in the model. MCDA has a relatively strong axiomatic basis.

Outranking methods, on the other hand, seek to make fewer assumptions about the nature of the underlying process that produces preferences.

It leaves more of the process of finalizing choice to the decision maker through fine-tuning in terms of items like the concordance and discordance thresholds. It recognizes the fact that options which record very poor relative performances on particular dimensions may be hard to implement in practice. It is a more interactive process between decision maker and model (Communities and Local Government 2009).

1.5.5 Fuzzy MCA

Fuzzy MCA methods are at the moment largely confined to the academic literature or to experimental applications, although ideas about MCA based on fuzzy sets have been discussed for more than 20 years. Fuzzy sets, conceptualized by Zadeh in the 1960s, are broadly equivalent to the sets found in conventional mathematics and probability theory with one important exception. The exception is that, instead of membership of a set being crisp (that is, an element is either definitely a member of a given set or it is not), set membership is graduated, or fuzzy or imprecise.

Set membership is defined by a membership function, $\mu(x)$, taking values between zero and one. Thus a particular issue might be regarded as a member of the set of major social concerns with a membership value of 0.8.

A membership function value of 0 conveys definitely not a member of the set, while $\mu = 1$ conveys definitely a member of the set. $\mu = 0.8$ suggests quite a strong degree of belief that the problem is a major one, but not complete certainty.

Proponents of fuzzy MCA would argue that one of the strengths of the fuzzy approach is that it recognizes the reality that many of the concepts involved in decision-making are far from clear or precise to those involved. Fuzzy sets provide an explicit way of representing that vagueness in the decision maker's mind in an explicit way. Developing this line of argument has led to many suggestions for fuzzy extensions to conventional MCA methods, such as fuzzy outranking methods and fuzzy utility theory (Chen and Hwang 1992). Nonetheless, it remains that, overall in the MCA community, enthusiasm for fuzzy MCA remains muted. Reasons for this include:

- A lack of convincing arguments that the imprecision captured through fuzzy sets and the mathematical operations that can be carried out on them actually match the real fuzziness of perceptions that humans typically exhibit in relation to the components of decision problems; (French 1988).
- Doubts as to whether prescriptively trying to model imprecision, which is in some sense a descriptive reflection of the failings of unaided human decision processing, is the right way to provide support to deliver better decisions;
- Failure to establish ways of calibrating membership functions and manipulating fuzzy values that have a transparent rationale from the point of view of non-specialists.

In combination, issues such as these continue to throw substantial doubt on the practical value of fuzzy MCA as a practical tool for supporting the main body of public decisions (Communities and Local Government 2009).

1.6 Conclusion

Multi-criteria decision making deals with qualitative criteria with uncertain information in order to express their decisions. It is common that the group of experts involved in such problems have different degrees of knowledge about the criteria.

The issues with MCDM were resolved using various methods. The following chapter identifies the different applications and use of the multicriteria approach in several fields.

References

Azzabi, L., Ayadi, D., Kobi, A., Robledo, C., & Boujelbenne, Y. (2009). Application of six sigma and Promethee multicriteria method to select the product system. *Quality Assurance, xv*(58), 11–17.

Bazerman, M. H. (1998). *Judgement in managerial decision making* (4th ed.). New York: Wiley.

Brownlow, S. A., & Watson, S. R. (1987). Structuring multi-attribute value hierarchies. *Journal of the Operational Research Society, 38*, 309–317.

Chen, S. J., & Hwang, C. L. (1992). *Fuzzy multiple attribute decision making: Methods and applications*. Berlin: Springer.

Communities and local government. (2009). *Multi-criteria analysis: A manual*. London: Department for Communities and Local Government.

E Costa, C. A., & Vansnick, J. C. (1994). MACBETH: An interactive path towards the construction of cardinal value functions. *International Transactions in Operational Research, 1*(4), 489–500. CP uses the Constraint Satisfaction Problem (CSP) (Mackworth 1977) framework to model combinatorial optimization problems.

Fishburn, P. C. (1965). Analysis of decisions with incomplete knowledge of probabilities. *Operations Research, 13*, 217–237.

French, S. (1988). *Decision theory: An introduction to the mathematics of rationality*. Chichester: Ellis Horwood.

Goodwin, P., & Wright, G. (1988). *Decision analysis for management Judgement*. Chichester: Wiley.

Howard, R. (1966). *Decision analysis: Introductory lectures on uncertainty*. Cambridge: Cambridge University Press.

Keeney, R. L., & Merkhofer, M. W. (1987). A multiattribute utility analysis of alternative sites for the disposal of nuclear waste. *Risk Analysis, 7*(2), 173–194.

Keeney, R. L., & Raiffa, H. (1976). *Decisions with multiple objectives: Preferences and value tradeoffs*. New York: Wiley. reprinted, Cambridge University Press, 1993.

Mousseau, V., Slowinski, R., & Zielniewicz, P. (2000). A user-oriented implementation of the ELECTRE TRI method integrating preference elicitation support. *Computers & Operations Research, 27*, 757–777.

Roy, B. (1998). A missing link in OR-DA: Robustness analysis. *Foundations of Computing and Decision Sciences, 23*, 141–160.

Roy, B. (1985). The optimization problem formulation: Criticism and overstepping. *Journal of the Operational Research Society, 32*(6), 427–436.

Roy, B. (1990). Decision-aid and decision-making. In *Readings in multiple criteria decision aid* (pp. 17–35). Berlin: Springer-Verlag.

Roy, B. (1974). Criteres multiples et modelisation des preferences: l'apport des relations de surclassement. *Revue d'Economie Politique, 84*, 1–144.

Von Neumann, J., & Morgenstern, O. (1947). *Theory of games and economic behaviour* (2nd ed.). Princeton: Princeton University Press.

Chapter 2
The Multi-Objective Optimization Problem with Fuzzy Goal Programming

2.1 Introduction

Many present-day problems are multi-objective in nature and their solution requires consideration of conflicting objectives. Usually, they have a number of potentially Pareto-optimal solutions. An extensive knowledge of the problem is required in discriminating between solutions, eliminating the unwanted ones and accepting the required solution(s) by a decision-making process.

It is well known that multi-objective optimization models have found a lot of important applications in decision-making problems such as in economics theory, management science, and engineering design. Because of these applications, many literatures have been published to study optimality conditions, duality theories, and topological properties of solutions of multi-objective optimization problems. In the case of optimization problems, the idea of regularizing a problem by adding a strongly convex term to the objective function is an option. The regularization technique proved to be an invaluable tool in the solution of ill-posed problems, and an enormous amount of work has been devoted to its study.

In this chapter, a multi-objective optimization problem formulation based on goal programming methods solves the multi-objective problem which can tackle relatively large test systems. This method is based on optimization of the most preferred objective and considering the other objectives as constraints.

Optimization of a single objective oversimplifies the pertinent objective function in some potential mathematical programming application situations. These two statements introduce the general topic of multi-objective programming (Blake and McCarl 1983). Multi-objective programming involves recognition that the decision maker is responding to multiple objectives. Generally, objectives are conflicting, so not all objectives can simultaneously arrive at their optimal levels. Multi-objective programming formally permits formulations where:

L. Azzabi et al., *The Multi-Criteria Approach for Decision Support*, International Series in Operations Research & Management Science 300, https://doi.org/10.1007/978-3-030-57262-4_2

(a) Solutions are generated which are as consistent as possible with target levels of goals;
(b) Solutions are identified which represent maximum utility across multiple objectives;
(c) Solution sets are developed which contain all non-dominated solutions.

Multiple objectives can involve such considerations as leisure, decreasing marginal utility of income, risk avoidance, preferences for hired labor, and satisfaction of desirable, but not obligatory, constraints.

An assumed utility function is used to choose appropriate solutions. Several fundamentally different utility function forms have been used in multi-objective models.

2.2 Multi-Objective Optimization

2.2.1 Principle

Multi-objective optimization is a natural extension of the traditional optimization of a single-objective function. If the multi-objective functions are commensurate, or non-competing, minimizing one objective function minimizes all criteria and using traditional optimization techniques can solve the problem (Baltes et al. 1994). However, if the objective functions are incommensurate, or competing, then minimization of one objective functions requires a compromise in another objective function. The competition between multi-objective functions gives rise to the distinguishing difference between multi-objective optimization and traditional single-objective optimization. This fact causes lack of complete order for multi-objective optimization problems (Biegler et al. 1986).

A multi-objective optimization problem usually has a set of Pareto-optimal solutions, instead of a single optimal solution. Thus, the goal of a multi-objective optimization is different from that of a single optimization. In multi-objective optimization, the goal is to find as many Pareto-optimal (or close to Pareto-optimal) solutions as possible. Since classical optimization methods only work with one solution at each iteration (Deb 1995), in order to find several Pareto-optimal solutions, they must be applied more than once, we hope to find a separate Pareto-optimal solution. (Fonseca and Fleming 1995; Horn et al. 1994; Srinivas and Deb 1995; Zitzler and Thiele 1998).

Before we discuss the problem features that may cause multi-objective difficulty, let us mention a couple of matters. First, we consider all objectives to be of minimization type. It is worth mentioning that identical properties as discussed here may also exist in problems with mixed optimization types (some are minimization and some are maximization). The concept of non-domination among solutions addresses only one type of problem.

2.2.2 Formulation Multi-Objective Problem

Consider a multi-objective optimization problem with p criteria or objectives where X denotes the finite set of feasible solutions. Each solution $x \in X$ is represented in the criterion space by its corresponding criterion Vector $f(x) = (f_1(x), \ldots \ldots, f_p(x))$. We assume in the following that each criterion has to be maximized. From these p criteria, the dominance relation defined on X, denoted by $\underline{\varnothing}$, states that a feasible solution x dominates a feasible solution x', $x \underline{\varnothing}\, x'$ if and only if $f_i(x) \geq f_i(x')$ for $i = 1, \ldots, p$. We denote by Δ the asymmetric part of $\underline{\varnothing}$. A solution x is efficient if and only if there is no other feasible solution $x' \in X$ such that $x' \Delta\, x$, and its corresponding criterion vector is said to be non-dominated. Thus, the efficient set is defined as $E(x) = \{x \in X : \forall\, x' \in X, (x' \Delta x)\}$ (Teng and Tzeng 1996).

The set of non-dominated criterion vectors, which corresponds to the image of the efficient set in the criterion space, is denoted by ND. Since the efficient set can contain different solutions corresponding to the same criterion vector, any subset of $E(x)$ that contains one and only one solution for every non-dominated criterion vector is called a reduced efficient set. Observe that $X' \subseteq X$ is a reduced efficient set if and only if it is a covering and independent set of X with respect to $\underline{\varnothing}$. We recall that, given $\succeq$ a binary relation defined on a finite set A,

- $B \subseteq A$ is a covering (or dominating) set of A with respect to $\succ$ if and only if for all $a \in \{A, B\}$, there exists $b \in B$ such that $b \succeq a$.
- $B \subseteq A$ is an independent (or stable) set with respect to $\succeq$ if and only if for all b, $b' \in B$, $b \neq b'$, not ($b \succeq b'$) (Ehrgott 2005).

2.3 Basic Structure of Fuzzy Goal Programming

2.3.1 Principle

Goal programming (GP) models were originally introduced by Charnes and Cooper in early 1961 for a linear model. This approach allows the simultaneous solution of a system of complex objectives and the solution of the problem requires the establishment among multiple objectives.

The main concept of Goal programming (GP) is to initially transform several goals into a specific numerical goal for each goal. The objective function is then formulated and a solution is sought which minimizes the weighted sum of the deviations from their respective objective. In general, GP models consist of three components: an objective function, a set of objective constraints, and non-negativity requirements. However, the target value associated with each goal may be unclear in the real world application. Fuzzy set theory is regularly used in recent research. A fuzzy set can be characterized by a membership function, which attributes to each object of a domain its membership rank (Zadeh 1965). The more an element or object can be considered to belong to a fuzzy set, the closer its membership degree is to 1. Different types of membership functions can be used to support the fuzzy analytical framework although the fuzzy description is hypothetical and that the values of belonging are subjective (Belmokaddem et al. 2009).

Membership functions, such as linear, piecewise linear, exponential, and hyperbolic functions, were used in different analysis. In general, the non-increasing and non-decreasing linear membership functions are frequently applied for the inequalities with less than or equal to and greater than or equal to relationships, respectively. Since the solution procedure of the fuzzy mathematical programming is to satisfy the fuzzy objective, a decision in a fuzzy environment is thus defined as the intersection of those membership functions corresponding to fuzzy objectives (Zimmermen 1978). Hence, the optimal decision could be any alternative in such a decision space that can maximize the minimum, represented by those corresponding membership functions (Zimmermen 1985).

The integrated use at GP and fuzzy sets theory has already been reported in the literature, (Hannan 1981a, b; Leberling 1981; Luhandjula 1982; Rubin and Narasimhan 1984; Tiwari et al. 1987; Wang and Fu 1997; Chen and Tsai 2001; Yaghoobi and Tamiz 2007) further integrated several fuzzy linear and multi-objective programming techniques.

2.3.2 *Goal Programming Problem*

The goal programming is a multi-objective mathematical programming model that attempts to produce an optimum solution for several conflicting objectives. The optimal solution would be the most satisfactory solution within a set of feasible solutions (Aouni and Kettani 2001). The GP model is a more flexible technique that is relatively easy to understand and utilize. There have been previous applications of GP in industrial and production settings; however, none of the early uses of GP incorporates explicitly the manager's preferences.

While we apply GP to capacity assessment problem we will consider preferences of the decision maker explicitly expressed through the concept of satisfaction function (Martel and Aouni 1990, 1996).

The overall purpose of GP is to minimize the deviations between the achievement of goals and their aspiration levels. A typical GP is expressed as follows:

$$\text{Minimize} \sum_{i=1}^{k} \left| F_i(x) - g_i \right| \tag{2.1}$$

$$\text{Subject to} x \in X = \left\{ x \in R^n; A_X \le b; x \ge 0 \right\} \tag{2.2}$$

where F_i the linear is function of the ith goal and g_i is the aspiration level of ith goal.

$$\text{Let } F_i(x) - g_i = d_i^+ - d_i^-; d_i^+, d_i^- \geq 0 \tag{2.3}$$

Equation (2.1) can be formulated as follows:

$$\text{Minimize} \sum_{i=1}^{k} \left| d_i^+ + d_i^- \right|$$

$$\text{Subject to} F_i(x) - d_i^+ + d_i^- - g_i = 0; i = 1, 2, \ldots\ldots. k$$

where $d_i^+ \geq 0$; $d_i^- \geq 0$ are, respectively, under deviations of ith goal.

2.3.3 Formulation Fuzzy Goal Programming

Narasimhan (1980) were the first to give a FGP formulation by using the concept of the membership functions. These functions are defined on the interval [0, 1]. So, the membership function for the i-th goal has a value of 1 when this goal is attained and the DM is totally satisfied; otherwise the membership function assumes a value between 0 and 1. Linear membership functions are used in literature and practice more than other types of membership functions. For the above three types of fuzzy goals linear membership functions are defined and depicted as follows (Fig. 2.1): where L_k (or u_k) is lower tolerance limit for kth fuzzy goal $G_k(x)$. They are either subjectively chosen by decision makers or tolerances in a technical process.

2.3.4 Fuzzy Linguistic for Determining the Degree of Achievement

The determination of a desirable achievement degree for a goal could be a difficult task for a decision-making in a fuzzy environment when using method by Chen and Tsai (2001). For assessing desirable achievement degrees imprecisely, a useful method is to use linguistic terms such as "Low Important," "Somewhat High Important," and "Very High Important," and so on to verbally describe the importance of each fuzzy goal. The associated membership functions are then defined. We can define $u_I(\alpha)$ to represent the membership function of each linguistic values about the importance of different objectives, where$u_I(\alpha) \in [0, 1]$, and α denotes the variable taking an achievement degree in the interval of $[\alpha_{\min}, \alpha_{\max}]$, $0 \leq \alpha_{\min} \leq \alpha_{\max} \leq 1$.

Then fuzzy numbers ranking methods can be used to map a membership function representing a fuzzy goal's importance to a real number in the range of [0, 1]. The real number obtained can be considered as the desirable achievement degree for the fuzzy goal.

Membership function	Analytical definition
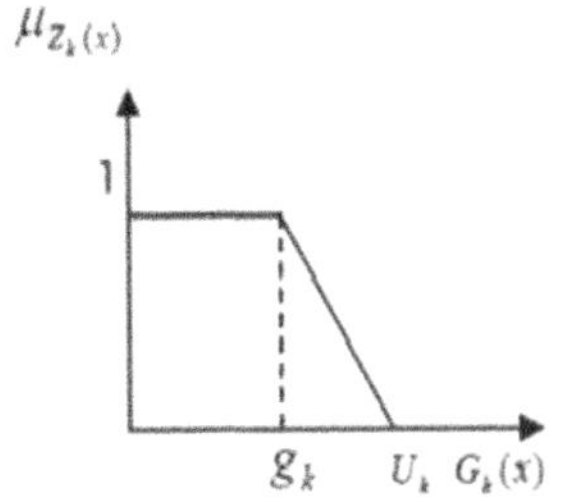	$\mu_{Z_k(x)} = \begin{cases} 1. & if..G_k(x) \le g_k \\ \dfrac{u_k - G_k(x)}{u_k - g_k} & .if..g_k \le G_k(x) \le u_k...k = 1,...,m...(1) \\ 0. & if..G_k(x) \ge \text{Uk} \end{cases}$
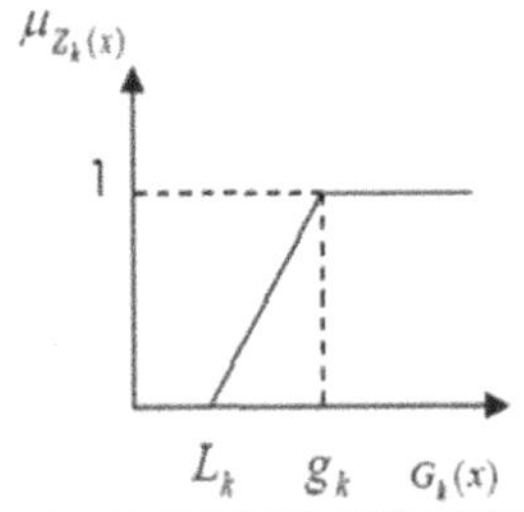	$\mu_{Z_k(x)} = \begin{cases} 1 & if..G_k(x) \ge g_k \\ \dfrac{G_k(x) - L_k}{g_k - L_k} & .if..L_k \le G_k(x) \le g_k...k = m+1,....,n....(2) \\ 0 & if..G_k(x) \le L_k \end{cases}$
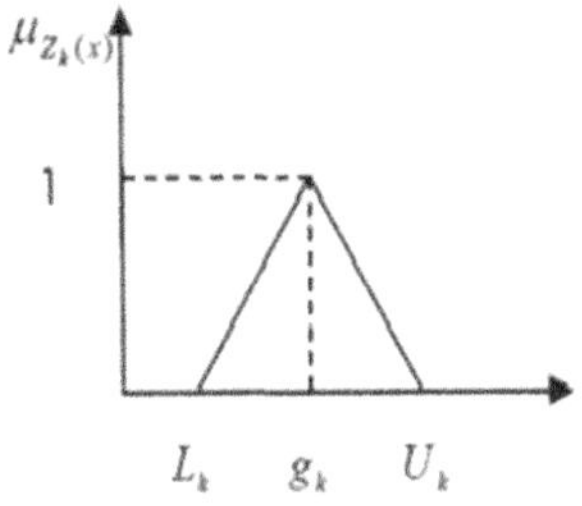	$\mu_{Z_k(x)} = \begin{cases} 0. & if..G_k(x) \le L_k \\ \dfrac{G_k(x) - L_k}{g_k - L_k} & .if..L_k \le G_k(x) \le g_k...k = n+1,....,l \\ \dfrac{u_k - G_k(x)}{u_k - g_k} & .if...g_k \le G_k(x) \le u_k \\ 0 & if..G_k(x) \ge u_k \end{cases}$

Fig. 2.1 Fuzzy goals linear membership functions

We define I = {Very Low Important = VLI, Low Important = LI, Somewhat Low Important = SLI, Medium = M, Somewhat High Important = SHI, High Important = HI, Very High Important = VHI}, as a set of linguistic values about the importance of different goals, (Fig. 2.2). shows the $u_i(\alpha)$ for these linguistic values. Triangular fuzzy numbers corresponding to these linguistic values are: VLI = (0,0,10%), LI = (5%,15%,25%), SLI = (20%, 32.5%, 45%), M = (40%, 50%, 60%), SHI = (55%, 67.5%, 80%), HI = (75%, 85%, 95%), VHI = (90%, 100%, 100%) (Fig. 2.2).

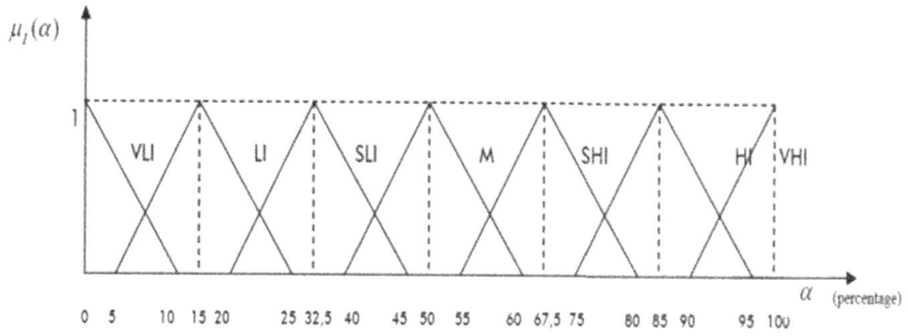

Fig. 2.2 Functions for linguistic values about the importance of different objectives

2.4 Numerical Example with Fuzzy Goal Programming

2.4.1 Data Description

In this section, we discuss the case of a joint shipping company, which seeks to maximize the profitability of its customer services and this by minimizing the number of road accidents for each vehicle (bus). A bus incurs three major networks: network (1) city university, network (2) town center and network, (3) industrial zone.

As the following Parameters and constants:

dit: Travel time of driver i for period j;

cit: Assure cost of the accident the driver i in period j;

rt: Regular time work force cost per employee hour in period t;

vit: Speed incurred by the driver i during period j;

Kit: Level of knowledge of road network driver i in period t;

Iio: number of accidents made by the drivers i in period t−1;

T: planning horizon paths of travel;

N: Total number of drivers;

Pit: number of conductor i in period t;

Iit: number of accidents made by the drivers i in period t;

Ht: worker hired in period t;

Ft: workers laid off in period t;

Iit. Min: minimum conductor i in period t;

Wt: Total number of buses during the period t;

WMin: Minimum number of buses available in period t;

WMax: Maximum number of buses available in period t;

The proposed model implementation in the company has the following conditions:

1. There is a six period planning horizon,
2. Three road networks,

3. Number minimum of accidents made by the drivers i in period t − 1: I10 = 18 accidents in the network 1, I20 = 10 accidents in the network 2, I30 = 33 accidents in the network 3.
4. Minimum number of accident during the period t of drivers i is 30 non-fatal accident.
5. Cost of repair or re-purchase bus accident per bus are, respectively, 10,386£ and 27,900£.
6. The linguistic values about the importance of objectives are: Very High Important = VHI, High Important = HI, Medium = M, respectively. And assumed that we have moderate decision maker, with $\alpha = 0.5$.
7. The cost of one driver in the bus of three networks during the t period is Rt = 2500£/man.
8. Minimum number of buses available in period t is W_{Min} = 50 bus of three networks.
9. Minimum number of buses available in period t is W_{Max} = 80 bus of three networks.
10. The initial number of bus WMin = 130 bus of three networks.
11. Number maximum of accidents made by the drivers i in period t of three networks is 80 accidents (Table 2.1).

Table 2.1 The basic data provided in the three networks

	Period	dit	vit	cit	Kit
Network 1	1	210	50 km/h	2500£	21
	2	120	40 km/h	2300£	23
	3	150	30 km/h	2450£	24
	4	120	50 km/h	1987£	26
	5	72	80 km/h	2365£	23
	6	138	40 km/h	2789£	24
Network 2	1	150	50 km/h	2500£	21
	2	120	50 km/h	1900£	25
	3	138	50 km/h	2450£	22
	4	210	60 km/h	1987£	21
	5	72	40 km/h	2365£	24
	6	150	50 km/h	1789£	25
Network 3	1	120	50 km/h	2500£	22
	2	150	70 km/h	2380£	24
	3	138	50 km/h	2450£	25
	4	210	50 km/h	1987£	26
	5	120	50 km/h	1865£	25
	6	72	80 km/h	2789£	21

2.4.2 Formulate and Solving Problem

– Construct the membership functions:

The linear membership function of each objective function is determined by asking the decision-making to specify the interval $[g_k \ldots u_k]$ of the objective values, and also to specify the equivalence of these objective values as a membership value in the interval [0, 1]. The linear and continuous membership function is found to be suitable for quantifying the fuzzy aspiration levels (Figs. 2.3, 2.4, and 2.5).

Then, the objective functions are:

$$\max f(u) = \sum_{k=1}^{3} u_k$$

Subject to:

$$u_1 \le \frac{3000 - Z_1}{7000}$$

$$u_2 \le \frac{80 - Z_2}{80}$$

$$u_3 \le \frac{90 - Z_3}{90}$$

$$P_{it} - K_{it} \times W_t \le 0$$
$$P_{it} + I_{i0} - I_{it} = d_{it}$$
$$W_t - W_0 - H_t + F_t \le W_t$$

Function 1: Minimize the total assure cost of the accident

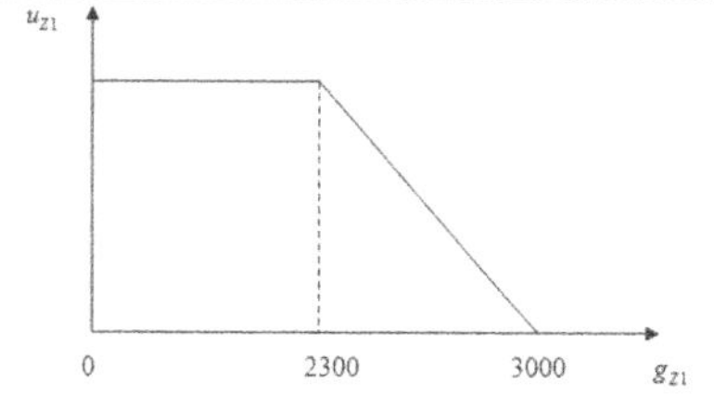

$$u_{z1} = \begin{cases} 1 & if;\ Z_1 \le 2300 \\ \frac{3000 - Z_1}{3000 - 2300}; & if\ 2300 \le Z_1 \le 3000 \\ 0. & if;\ Z_1 \ge 3000 \end{cases}$$

Fig. 2.3 Membership function Z_1

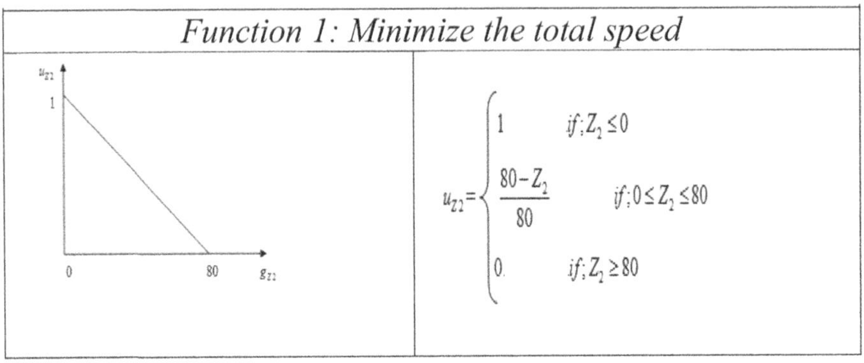

Fig. 2.4 Membership function Z_2

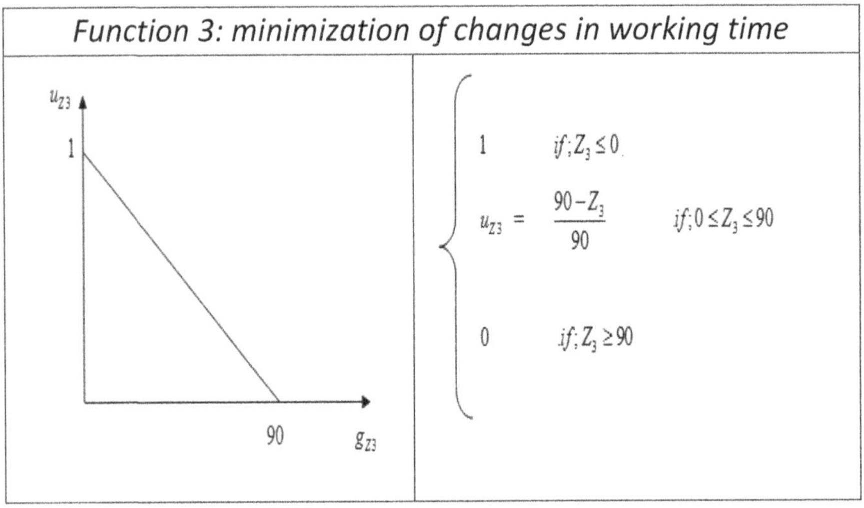

Fig. 2.5 Membership function Z_3

$$W_{\min} \leq W_t \leq W_{\max}$$

$$\sum_{i=1}^{3} I_{it} \leq 61$$

$$I_{it} \geq 30$$

$I_{10} = 18; I_{20}=10; I_{30}=33$

$$u_1 \geq 0.72$$
$$u_2 \geq 0.85$$
$$u_3 \geq 0.50$$

$$I_{it}, u_1, u_2, u_3 \geq 0$$
$$i = 1,2,3$$
$$t = 1,2,3,4,5,6$$

- Solve problem:

  The LINGO computer software package was used to run the programming model (Table 2.2):

  Using fuzzy goal programming to simultaneously minimize total road accident costs (Z_1), carrying costs (Z_2), and changes in Work force levels (Z_3), yields total road accident cost of 2999 £, and resulting achievement degrees for the three fuzzy goal (u_1,$u_2$and u_3) are 0, 0, and 1 respectively, all of which satisfy the requirements of decision makers.

2.4.3 Conclusion

In this chapter, we presented a road accident constraint in which data of three networks are fuzzy values and the objective function assumes multiple objectives. Then, non-linear constraints are linearized by defining and adding auxiliary

Table 2.2 Optimal solution model

Period	Networks	I_{it}	W_{it}	P_{it}
0	1	18	80	–
	2	10		
	3	33		
1	1	0	80	55
	2	0		
	3	30		
2	1	18	80	55
	2	0		
	3	10		
3	1	0	80	55
	2	0		
	3	30		
4	1	0	80	55
	2	10		
	3	18		
5	1	0	80	55
	2	0		
	3	33		
6	1	0	80	55
	2	0		
	3	30		

constraints. Finally, the optimal solution of the problem is founded by solving the linear programming problem with fuzzy and crisp constraints and applying fuzzy programming technique.

Various fuzzy approaches have been proposed for the solution of multi-objective linear plus linear fractional programming problem and most of the approaches have computational burdensome. Our approach is to give simple procedure for the solution of multi-objective fuzzy goal programming problem.

References

Aouni, B., & Kettani, O. (2001). Goal programming model: A glorious history and a promising future. *European Journal of Operational Research, 133*(2), 1–7.

Baltes, M., Schneider, R., Sturm, C., & Reuss, M. (1994). Optimal experimental design for parameter estimation in unstructured growth models. *Biotechnology Progress, 10*, 480–488.

Belmokaddem, M., Mekidiche, M., & Sahed, A. (2009). Application of a fuzzy GOAL PROGRAMMING approach with different importance and priorities to aggregate production planning. *Journal of Applied Quantitative Methods, 4*(3).

Biegler, L. T., Damiano, J. J., & Blau, G. E. (1986). Nonlinear parameter estimation: A case study comparison. *A.I.Ch.E. Journal, 32*, 29–45.

Blake, B. F., & McCarl, B. A. (1983). Goal programming via multidimensional scaling applied to Senegalese subsistence farming: A reply. *American Journal of Agricultural Economics, 65*, 832–833.

Charnes, A. and Cooper, W. (1961) Management models and industrial applications of linear programming, Wiley, New York.

Chen, L. H., & Tsai, F. C. (2001). Fuzzy goal programming with different importance and priorities. *European Journal of Operational Research, 133*, 548–556.

Deb, K. (1995). *Optimization for engineering design: Algorithms and examples*. New Delhi: Prentice-Hall.

Ehrgott, M. (2005). Multicriteria optimization. In *LNEMS 491*. Berlin: Springer.

Fonseca, C. M., & Fleming, P. J. (1995). An overview of evolutionary algorithms in multi-objective optimization. *Evolutionary Computation, 3*, 1–16.

Hannan, E. L. (1981a). Linear programming with multiple fuzzy goals. *Fuzzy Sets and Systems, 6*, 235–248.

Hannan, E. L. (1981b). On Fuzzy Goal Programming. *Decision Sciences, 12*, 522–531.

Horn, J., Nafploitis, N., & Goldberg, D. E. (1994). A niched Pareto genetic algorithm for multi-objective optimization. In Z. Michalewicz (Ed.), *Proceedings of the first IEEE conference on evolutionary computation* (pp. 82–87). Piscataway, NJ: IEEE Service Center.

Leberling, H. (1981). On finding compromise solutions in multi criteria problems using the fuzzy min-operator. *Fuzzy Sets and Systems, 6*, 105–118.

Luhandjula, M. K. (1982). Compensatory operations in fuzzy programming with multiple objectives. *Fuzzy Sets and Systems, 8*, 245–252.

Martel, J. M., & Aouni, B. (1996). Incorporating the decision-maker's preferences in the goal programming model with fuzzy goals values: A new formulation. In *Lecture notes in economics and mathematical systems*. Berlin: Springer-Verlag.

Martel, J.-M., & Aouni, B. (1990). Incorporating the decision-maker's preferences in the goal programming model. *Journal of Operational Research Society, 41*, 1121–1132.

Narasimhan, R., (1980). *Goal programming in a fuzzy environment, Decision Sciences, 11*(2): 259–410.

Rubin, P. A., & Narasimhan, R. (1984). Fuzzy goal programming with nested priorities. *Fuzzy Sets and Systems, 14*, 115–129.
Srinivas, N., & Deb, K. (1995). Multi-objective function optimization using non-dominated sorting genetic algorithms. *Evolutionary Computation, 2*(3), 221–248.
Teng, J., & Tzeng, G. (1996). A multi-objective programming approach for selecting non-independent transportation investment alternatives. *Transportation Research-B, 30*(4), 201–307.
Tiwari, R. N., Dharmar, S., & Rao, J. R. (1987). Fuzzy goal programming – An additive model. *Fuzzy Sets and Systems, 24*, 27–34.
Wang, H. F., & Fu, C. C. (1997). A generalization of fuzzy goal programming with preemptive structure. *Computers and Operations Research, 24*, 819–828.
Yaghoobi, M. A., & Tamiz, M. (2007). A method for solving fuzzy goal programming problems based on MINMAX approach. *European Journal of Operational Research, 177*, 1580–1590.
Zadeh, L. A. (1965). Fuzzy sets. *Information and Control, 8*, 338–353.
Zimmermen, H. J. (1978). Fuzzy programming and linear programming with several objective functions. *Fuzzy Sets and Systems, 1*, 45–56.
Zimmermen, H. J. (1985). Applications of fuzzy sets theory to mathematical programming. *Information Science, 35*, 29–58.
Zitzler, E., & Thiele, L. (1998). Multi-objective optimization using evolutionary algorithms—A comparative case study. In A. E. Eiben, T. Bäck, M. Schoenauer, & H.-P. Schwefel (Eds.), *Parallel problem solving from nature* (Vol. V, pp. 292–301). Berlin: Springer.

Chapter 3
Fuzzy Modes and Effects Analysis by Using Fuzzy AHP

3.1 Introduction

Failure Mode and Effect Analysis (FMEA) is a systematic process to identify potential design or process failures before they occur. The objective is to eliminate them or to minimize the risk associated with them. The method is a procedure for analyzing failure modes and classifying them by severity. It is a systematic process for identifying potential failures before they occur with the aim of eliminating them or minimizing the risk associated with them. The percentages of failures are identified by a group of experts (Ravi and Prabhu 2001). The FMEA, which provides a framework for the analysis of the causes and effects of potential product failures (Chin et al. 2008), aims to prioritize the risk priority number (RPN) of the design or planning process of the product. product to allocate limited resources.

FMEA, designed to provide information for risk management decision-making (Pillay and Wang 2003), was first developed as a formal design methodology by NASA in 1963 for their obvious reliability requirements and then, it was adopted and promoted by Ford Motor in 1977 (Chin et al. 2008). Since then, it has become a powerful tool extensively used for safety and reliability analysis of products and processes in a wide range of industries, especially aerospace, nuclear, and automotive industries (Gilchrist 1993; Sharma et al. 2005).

A typical FMEA is consisted of the following components; the identification and listing of failure modes and the consequent faults, assessment of the chances of the occurrence of faults, then assessment of the chances of the detection of faults, assessment of the severity of the consequences of the faults, calculation of a measure of the risk, the ranking of the faults based on the risk, taking action on the high-risk problems, and checking the effectiveness of the action with the use of a revised risk measurement (Ben and Raouf 1996). Each failure mode can be evaluated

L. Azzabi et al., *The Multi-Criteria Approach for Decision Support*, International Series in Operations Research & Management Science 300,
https://doi.org/10.1007/978-3-030-57262-4_3

by three factors as severity, likelihood of occurrence, and the difficulty of detection of the failure mode. In a typical FMEA evaluation, a number between 1 and 10 (with 1 being the best and 10 being the worst case) is given for each of the three factors. By multiplying the values for severity (S), occurrence (O), and detectability (D), a risk priority number (RPN) is obtained, which is RPN = S. O. D (Can Kutlu and Ekmekcioğlu 2012). Then the RPN value for each failure mode is ranked to find out the failures with higher risks.

3.2 The Traditional FMEA

In order to analyze a specific product or system, a cross-functional team should be established for carrying out FMEA. The members can be from the fields of design, manufacturing, management, service, etc. The first step in FMEA is to identify all possible potential failure modes of the product or system. After that, critical analysis is performed on these failure modes considering the following three risk factors: O, S, and D. Generally, risk degree of a failure can be represented by a RPN that is defined as the product of the scores of O, S, and D: RPN = O. S. D.

FMEA has been proven to be one of the most important early preventative initiatives during the design stage of a system, product, process, or service (Bowles 2004). However, the RPN has been extensively criticized for various reasons:

- Different sets of O, S, and D ratings may produce exactly the same value of RPN, but their hidden risk implications may be totally different. For example, two different events with values of 2, 3, 2 and 4, 1, 3 for O, S, and D, respectively, will have the same RPN value of 12. However, the hidden risk implications of the two events may be very different because of the different severities of the failure consequence (Sankar and Prabhu 2001). This may cause a waste of resources and time, or in some cases, a high-risk event being unnoticed.
- The relative importance among O, S, and D is not taken into consideration. The three factors are assumed to have the same importance. This may not be the case when considering a practical application of FMEA.
- The mathematical formula for calculating RPN is questionable and debatable. There is no rationale as to why O, S, and D should be multiplied to produce the RPN.
- The conversion of scores is different for the three factors. For example, a linear conversion is used for O, but a non-linear transformation is employed for D.
- RPNs are not continuous with many holes and heavily distributed at the bottom of the scale from 1 to 1000. This causes problems in interpreting the meaning of the differences between different RPNs. For example, is the difference between the neighboring RPNs of 1and 2 the same or less than the difference between 900 and 1000?
- The RPN considers only three factors mainly in terms of safety. Other important factors such as economic aspects are ignored.

- Small variations in one rating may lead to vastly different effects on the RPN, depending on the values of the other factors. For example, if O and D are both 10, then a 1-point difference in severity rating results in a 100-point difference in the RPN; if O and D are equal to 1,then the same 1-point difference results in only a 1-point difference in the RPN; if O and D are both 4, then a 1-point difference produces a 16-point difference in the RPN.
- The three factors are difficult to precisely determine. Much information in FMEA can be expressed in a linguistic way such as likely, important, or very high and soon (Ben and Raouf 1996) (Tables 3.1, 3.2, and 3.3).

In traditional FMEA approach, the diversity and ability of the team are the most important considerations, followed by training for the team members. This leads to a high cost.

Furthermore, industrial practitioners usually find it hard to share their experience among team members of different background. This indeed prohibits the application of FMEA in a broader scope (Pillay and Wang 2003). Many decision-making and problem-solving tasks are too complex to be understood quantitatively; however, people succeed by using knowledge that is imprecise rather than precise.

3.3 The Fuzzy FMEA

The studies about FMEA considering fuzzy approach use the experts who describe the risk factors O, S, and D by using the fuzzy linguistic terms. The linguistic variables were used for evaluating three risk factors O, S, and D as an interpretation of the traditional 10-point scale (1–10) FMEA factor scores. In the fuzzy FMEA literature, the studies have mostly concerned with the fuzzy rule-based approach by using if-then rules (Bowles and Pelaez 1995; Chin et al. 2008; Guimaraes and Lapa 2004, 2007; Pillay and Wang 2003; Sharma et al. 2005; Tay and Lim 2006). After the assignments of the linguistic terms to the factors, if-then rules were generated taking the linguistic variables as inputs to evaluate the risks. The outputs of the fuzzy infer-

Table 3.1 Traditional ratings for occurrence of a failure

Rating	Probability of occurrence	Possible failure rate
10	Very high: Failure is almost inevitable	≥1/2
9		1/3
8	High: Repeated failures	1/8
7		1/20
6	Moderate: Occasional failures	1/80
5		1/400
4		1/2000
3	Low: Relatively few failures	1/15,000
2		1/150,000
1	Remote: Failure is unlikely	≤1/1,500,000

Table 3.2 Traditional ratings for severity of a failure

Rating	Effect	Severity of effect
10	Hazardous without warning	Very high severity ranking when a potential failure mode affects safe vehicle operation and/or involves noncompliance with government regulations without warning
9	Hazardous with warning	Very high severity ranking when a potential failure mode affects safe vehicle operation and/or involves noncompliance with government regulations with warning
8	Very high	Vehicle/item inoperable, with loss of primary function
7	High	Vehicle/item operable, but at reduced level of performance. Customer dissatisfied
6	Moderate	Vehicle/item operable, but comfort/convenience item(s) inoperable. Customer experiences discomfort
5	Low	Vehicle/item operable, but comfort/convenience item(s) operable at reduced level of performance.
		Customer experiences some dissatisfaction
4	Very low	Cosmetic defect in finish, fit and finish/squeak, or rattle item that does not conform to specifications.
		Defect noticed by most customers
3	Minor	Cosmetic defect in finish, fit and finish/squeak, or rattle item that does not conform to specifications.
		Defect noticed by average customer
2	Very minor	Cosmetic defect in finish, fit and finish/squeak, or rattle item that does not conform to specifications.
		Defect noticed by discriminating customers
1	None	No effect

ence system were variously named as risk, the critically failure mode (Xu et al. 2002), priority for attention (Pillay and Wang 2003), and fuzzy RPN (Sharma et al. 2005; Xu et al. 2002) in the fuzzy FMEA studies which consider the fuzzy rule-based approach. (Braglia and Bevilacqua 2000) drew attention to the doubts remained due to the difficulties in defining many rules and membership functions required by this methodology considering the applicability of the real industrial cases.

A conventional form of FMEA includes the design function of parts, the potential failure mode (categories of failure), the potential effects of failure (measured by the severity index), the potential causes of failure (measured by the occurrence (frequency) index), the detection method (measured by the detectability index), and the risk priority number (RPN). The RPN is used to evaluate the risk level of a part's failure mode in design stage, and is determined by the multiplication of three characteristic failure mode indexes, the severity of the potential failure (S), the frequency of potential failure (O), and the detectability index (D), respectively, as RPN = S.O.D. (Liang-hsuan and When-chang 2009).

Severity, Occurrence, and Detect, as in the traditional RPN function, are also used as the input factors for the fuzzy RPN function. The membership functions of these three factors are determined by interpreting the linguistic terms. Tables 3.4, 3.5, and 3.6:

Table 3.3 Traditional ratings for detection

Rating	Detection	Criteria
10	Absolutely impossible	Design control will not and/or cannot detect a potential cause/ mechanism and subsequent failure mode; or there is no design control
9	Very remote	Very remote chance the design control will detect a potential cause/ mechanism and subsequent failure mode
8	Remote	Remote chance the design control will detect a potential cause/ mechanism and subsequent failure mode
7	Very low	Very low chance the design control will detect a potential cause/ mechanism and subsequent failure mode
6	Low	Low chance the design control will detect a potential cause/ mechanism and subsequent failure mode
5	Moderate	Moderate chance the design control will detect a potential cause/ mechanism and subsequent failure mode
4	Moderately high	Moderately high chance the design control will detect a potential cause/mechanism and subsequent failure mode
3	High	High chance the design control will detect a potential cause/ mechanism and subsequent failure mode
2	Very high	Very high chance the design control will detect a potential cause/ mechanism and subsequent failure mode
1	Almost certain	Design control will almost certainly detect a potential cause/ mechanism and subsequent failure mode

Table 3.4 Scales for occurrence

Frequency of occurrence	Rating	Possible failure rate
Remote	1	<1:20,000
Low	2	1:20,000
	3	1:10,000
Moderate	4	1:2000
	5	1:1000
	6	1:200
High	7	1:100
	8	1:20
Very high	9	1:10
	10	1:2

3.4 The Fuzzy AHP

3.4.1 Principle

Analytic Hierarchy Process (AHP) is a powerful method to solve complex decision problems. Any complex problem can be decomposed into several sub-problems using AHP in terms of hierarchical levels where each level represents a set of criteria or attributes relative to each sub-problem. The AHP method is a multi-criteria method of analysis based on an additive weighting process, in which several relevant attributes are represented through their relative importance. From the AHP

Table 3.5 Scales for severity

Severity	Rating
Remote	1
Low	2
	3
Moderate	4
	5
	6
High	7
	8
Very high	9
	10

Table 3.6 Scales for detection

Detectability	Rating	Probability of detectability (%)
Remote	1	86–100
Low	2	76–85
	3	66–75
Moderate	4	56–65
	5	46–55
	6	36–45
High	7	26–35
	8	16–25
Very high	9	6–15
	10	0–5

method, multiple attributes are obtained from a pairwise comparison process, in which the relevance of attributes or categories of intangible asset factors is matched two against two in a hierarchical structure. However, the pure AHP model has some shortcomings (Yang and Chen 2004). They pointed out that the AHP method is mainly used in nearly crisp-information decision applications; the AHP method creates and deals with a very unbalanced scale of judgment; the AHP method does not take into account the uncertainty associated with the mapping of human judgment to a number by natural language; the ranking of the AHP method is rather imprecise; and the subjective judgment by perception, evaluation, improvement, and selection based on preference of decision makers have great influence on the AHP results (Ahmad et al. 2001). To overcome these problems, several researchers integrate fuzzy theory with AHP to improve the uncertainty. Hence, Buckley (1985) used the evolutionary algorithm to calculate the weights with the trapezoidal fuzzy numbers.

The Fuzzy Analytic Hierarchy Process Method is an Analytical Hierarchy Process (AHP) method developed with fuzzy logical theory. The Fuzzy AHP method is used similarly to the AHP method. It's just that the Fuzzy AHP method sets the AHP scale to the scale of the fuzzy triangle to be accessed first (Azamathulla and Wu 2011).

3.4.2 Determining the Evaluation Dimensions Weights

This research employs Fuzzy AHP to fuzzy hierarchical analysis by allowing fuzzy numbers for the pair-wise comparisons and finds the fuzzy preference-weights. In this section, we briefly review concepts for fuzzy hierarchical evaluation. Then the following sections will introduce the computational process about Fuzzy AHP in detail.

Then we will briefly introduce that how to carry out the fuzzy AHP in the following sections.

Step 1: Construct pair-wise comparison matrices among all the elements/criteria in the dimensions of the hierarchy system.

Assign linguistic terms to the pair-wise comparisons by asking which is the more important of each two dimensions, as following matrix Ã

$$\tilde{A} = \begin{bmatrix} 1 & \tilde{a}_{12} & \cdots & \tilde{a}_{1n} \\ \tilde{a}_{21} & 1 & \cdots & \tilde{a}_{2n} \\ \vdots & \vdots & \ddots & \vdots \\ \tilde{a}_{n1} & \tilde{a}_{n2} & \cdots & 1 \end{bmatrix} = \begin{bmatrix} 1 & \tilde{a}_{12} & \cdots & \tilde{a}_{1n} \\ 1/\tilde{a}_{12} & 1 & \cdots & \tilde{a}_{2n} \\ \vdots & \vdots & \ddots & \vdots \\ 1/\tilde{a}_{1n} & 1/\tilde{a}_{2n} & \cdots & 1 \end{bmatrix}$$

where

$$\tilde{a}_{ij} = \begin{cases} \tilde{9}-1,\tilde{8}-1,\tilde{7}-1,\tilde{6}-1,\tilde{5}-1,\tilde{4}-1,\tilde{3}-1,\tilde{2}-1,\tilde{1}-1,\tilde{1},\tilde{2},\tilde{3},\tilde{4},\tilde{5},\tilde{6},\tilde{7},\tilde{8},\tilde{9}, & i \neq j \\ 1, & i = j \end{cases}$$

Use the geometric mean technique to define the fuzzy geometric mean and the fuzzy weights of each criterion by Hsieh et al. (2004).

$$\tilde{r}_i = \left(\tilde{a}_{i1} \otimes \cdots \otimes \tilde{a}_{ij} \otimes \cdots \otimes \tilde{a}_{in}\right)^{1/n}$$

$$\tilde{w}_i = \tilde{r}_i \otimes \left[\tilde{r}_1 \oplus \cdots \oplus \tilde{r}_i \oplus \cdots \oplus \tilde{r}_n\right]^{-1}$$

where,is fuzzy comparison value of dimension i to criterion j, thus, is a geometric mean of fuzzy comparison value of criterion i to each criterion, w˜ i is the fuzzy weight of the ith criterion, can be indicated by a TFN, w˜ i = (lwi, mwi, uwi). The lwi, mwi, and uwi stand for the lower, middle, and upper values of the fuzzy weight of the ith dimension.

The fuzzy analytic hierarchy process (FAHP) approach was considered by Hua et al. (2009) for evaluating the relative weightings of the risk factors of FMEA to analyze of the risks of green components in compliance with the European Union (EU) the Restriction of Hazardous Substance (RoHS) directive in the incoming quality control (IQC) stage. In the study, Severity factor was explained by two criteria and with considering the occurrence and the detection factors, the FAHP was utilized to determine the weights of four criteria by two experts. The traditional

FMEA was modified to form green component risk priority number (GCRPN) for the calculation of the risks with regard to each category of green components. GCRPN was formulated by the sum of the terms of products of the factor scores and weights. Braglia and Bevilacqua (2000) proposed a fuzzy TOPSIS approach for Failure Mode, Effects and Criticality Analysis (FMECA). The fuzzy version of TOPSIS was applied allowing the traditional FMECA factors O, S, and D and their equally important weights to be evaluated using triangular fuzzy numbers.

3.5 Conclusion

Failure mode and effects analysis (FMEA) is a powerful tool for identifying and assessing potential failures. The tool has become increasingly important in new product development, manufacture, or engineering applications. The FMEA team often demonstrates different opinions and knowledge from one team member to another and produces different types of assessment information such as complete and incomplete, precise and imprecise, and known and unknown because of its cross-functional and multidisciplinary nature. These different types of information are very difficult to incorporate into the FMEA by the traditional risk priority number (RPN) model and fuzzy rule-based approximate reasoning methodologies. In this chapter we presented an FMEA using the fuzzy approach, a newly developed methodology for multiple attribute decision analysis. The proposed fuzzy FMEA is based to attribute the fuzzy analytical hierarchy process. As is illustrated by the numerical example in treatment gas industry.

References

Ahmad, Z., Azamathulla, H. M., & Ghani, A. A. (2001). ANFIS-based approach for estimation transverse mixing coefficient. *IWA-Water Science & Technology, 63*(5), 1004–1009.

Azamathulla, H. M., & Wu, F.-C. (2011). Support vector machine approach to for longitudinal dispersion coefficients in streams. *Applied Soft Computing, 11*(2), 2902–2905.

Ben-Daya, M., & Raouf, A. (1996). A revised failure mode and effects analysis model. *International Journal of Quality & Reliability Management, 13*(1), 43–47.

Bowles, J. B., & Pelaez, C. E. (1995). Fuzzy logic prioritization of failures in a system failure mode, effects and criticality analysis. *Reliability Engineering and System Safety, 50*, 203–213.

Bowles, J. B. (2004). An assessment of PRN prioritization in a failure modes effects and criticality analysis. *Journal of the IEST, 47*, 51–56.

Braglia, M., & Bevilacqua, M. (2000). Fuzzy modeling and analytic hierarchy processing as a means to quantify risk levels associated with failure modes in production systems. *Technology, Law and Insurance, 5*(3–4), 125–134.

Buckley, J. J. (1985). Fuzzy hierarchical analysis. *Fuzzy Sets and Systems, 17*(3), 233–247.

Can Kutlu, A., & Ekmekcioğlu, M. (2012). Fuzzy failure modes and effects analysis by using fuzzy TOPSIS-based fuzzy AHP. *Expert Systems with Applications, 39*(2012), 61–67.

Chin, K. S., Chan, A., & Yang, J. B. (2008). Development of a fuzzy FMEA based product design system. *International Journal of Advanced Manufacturing Technology, 36*, 633–649.

Gilchrist, W. (1993). Modeling failure mode and effect analysis. *International Journal of Quality and Reliability Management, 10*(5), 16–23.

Guimaraes, A. C. F., & Lapa, C. M. F. (2004). Fuzzy FMEA applied to PWR chemical and volume control system. *Progress in Nuclear Energy, 44*(3), 191–213.

Guimaraes, A. C. F., & Lapa, C. M. F. (2007). Fuzzy inference to risk assessment on nuclear engineering systems. *Applied Soft Computing, 7*, 17–28.

Hsieh, T. Y., Tzeng, S. T., & Lu, G. H. (2004). Fuzzy MCDM approach for planning and design tenders selection in public office buildings. *International Journal of Project Management, 22*(7), 573–584.

Hua, Z. S., Wei, Y., Dun, Z., Sai, L. N., Chong-Fa, C., A-Lin, L., & Lu Lia (2009). A fuzzy analytic hierarchy process (FAHP) approach to eco-environmental vulnerability assessment for the danjiangkou reservoir area, *China, Ecological Modelling, 220*(23), 10.

Liang-hsuan, C., & When-chang, K. (2009). Fuzzy linear programming models for new product design using QFD with FMEA. *Applied Mathematical Modelling, 33*, 633–647.

Pillay, A., & Wang, J. (2003). Modified failure mode and effects analysis using approximate reasoning. *Reliability Engineering & System Safety, 79*, 69–85.

Ravi, N., & Prabhu, B. S. (2001). Application of fuzzy logic to matrix FMECA. In D. O. Thompson & D. E. Chimenti (Eds.), *Review of Progress in quantitative nondestructive evaluation* (Vol. 20). New York: American Institute of Physics.

Sankar, N. R., & Prabhu, B. S. (2001). Modified approach for prioritization of failures in a system failure mode and effects analysis. *International Journal of Quality & Reliability Management 2001, 18*(3), 324–335.

Sharma, R. K., Kumar, D., & Kumar, P. (2005). Systematic failure mode effect analysis (FMEA) using fuzzy linguistic modeling. *International Journal of Quality and Reliability Management, 22*(9), 986–1004.

Tay, K. M., & Lim, C. P. (2006). Fuzzy FMEA with a guided rules reduction system for prioritization of failures. *International Journal of Quality and Reliability Management, 23*(8), 1047–1066.

Xu, K., Tang, L. C., Xie, M., Ho, S. L., & Zhu, M. L. (2002). Fuzzy assessment of FMEA for engine systems. *Reliability Engineering and System Safety, 75*, 17–29.

Yang, C. C., & Chen, B. S. (2004). Key quality performance evaluation using fuzzy AHP. *Journal of the Chinese Institute of Industrial Engineers, 21*(6), 543–550.

Chapter 4
Multi-Criteria Approach to Classification Systems of Control Aircraft Flight: Application PROMETHEE Method

4.1 Introduction

The multi-Criteria decision aid (MCDA) has been one of the very fast growing areas of Operational Research (OR) during the two last decades. The MCDA often deals with ranking of many concrete alternatives from the best to the worst ones based on multiple conflicting criteria. The MCDA is also concerned with theory and methodology that can treat complex problems encountered in management, business, engineering, science, and other areas of human activity.

In recent years, several MCDA methods have been proposed to help in selecting the best compromise alternatives. The development of MCDA methods has been motivated not only by a variety of real-life problems requiring the consideration of multiple criteria, but also by practitioners' desire to propose enhanced decision-making techniques using recent advancements in mathematical optimization, scientific computing, and computer technology (Wiecek et al. 2008). The PROMETHEE method (Preference Ranking Organization Method for Enrichment Evaluations) is one of the most recent MCDA methods that was developed by Brans et al. (1986) and further extended by Vincke and Brans (1985).

4.2 Multi-Criteria Decision-Making: A General Overview

Decision problems present a number of decision criteria. These problems are often referred to as multi-criteria decision making (MCDM) (Al-Shemmeri et al. 1997).

Some of the industrial engineering applications of MCDM include the use of decision analysis in integrated manufacturing (Al-Kloub and Abu-Taleb 1998), in the evaluation of technology investment decisions (Al-Shemmeri et al. 1997), in

L. Azzabi et al., *The Multi-Criteria Approach for Decision Support*, International Series in Operations Research & Management Science 300, https://doi.org/10.1007/978-3-030-57262-4_4

flexible manufacturing systems (Bouyssou and Perny 1992), layout design (Doumpos and Zopounidis 2004), and also in other engineering problems.

Multi-criteria decision-making (MCDM) plays a critical role in many real-life problems. It is not an exaggeration to argue that almost any local or federal government, industry, or business activity involves, in one way or the other, the evaluation of a set of alternatives in terms of a set of decision criteria. Very often these criteria are conflicting with each other. Even more often the pertinent data are very expensive to collect (Bana e Costa 1990).

Multi-criteria decision-making (MCDM) is one of the most widely used decision methodologies in science, business, and engineering worlds (Chareonsuk et al. 1997). MCDM methods aim at improving the quality of decisions by making the process more explicit, rational, and efficient (Yang and Sen 1997; Yang and Xu 2002; Yan et al. 2007).

4.3 PROMETHEE Methods

The PROMETHEE family of outranking methods, including the PROMETHEE I for partial ranking of the alternatives and the PROMETHEE II for complete ranking of the alternatives, was developed by Brans and presented for the first time in 1982 (Brans 1982).

A few years later, several versions of the PROMETHEE methods such as the PROMETHEE III for ranking based on interval, the PROMETHEE IV for complete or partial ranking of the alternatives when the set of viable solutions is continuous, the PROMETHEE V for problems with segmentation constraints (Brans and Mareschal 1992).

The PROMETHEE VI for the human brain representation (Brans and Mareschal 1995) and the visual interactive module GAIA (Geometrical Analysis for Interactive Aid) for graphical representation (Brans and Mareschal 1994a) were developed to help in more complicated decision-making situations (Brans and Mareschal 2005). Figueira et al. (2004) have recently proposed two extended approaches on PROMETHEE, called as the PROMETHEE TRI for dealing with sorting problems and the PROMETHEE CLUSTER for nominal classification. The methods of PROMETHEE have successfully been applied in many fields and a number of researchers have used them in decision-making problems. The PROMETHEE methods have some requisites of an appropriate multi-criteria method and their success is basically due to their mathematical properties and to their particular friendliness of use (Brans and Mareschal 2005).

PROMETHEE is also a quite simple ranking method in conception and application compared with the other methods for multi-criteria analysis (Brans et al. 1986). Therefore, the number of practitioners who are applying the PROMETHEE method to practical multiple-criteria decision problems, and researchers who are interested in sensitivity aspects of the PROMETHEE method, increases year by year as can be illustrated by increasing numbers of scholarly papers and conference presentations.

The evaluation table is the starting point of the PROMETHEE method. In this table, the alternatives are evaluated on the different criteria. These evaluations involve essentially numerical data. The implementation of PROMETHEE requires two additional types of information, namely:

- Information on the relative importance of the criteria considered;
- Information on the decision makers preference function, which he/she uses when comparing the contribution of the alternatives in terms of each separate criterion.

The above information determines the preference structure of the decision maker.

- The weights: The weights can be determined according to various methods (Chou et al. 2004; Elevli and Demirci 2004). PROMETHEE does not provide specific guidelines for determining these weights, but assumes that the decision maker is able to weigh the criteria appropriately, at least when the number of criteria is not too large (Van Huylenbroeck 1995).
- The preference function: The preference function (P_j) translates the difference between the evaluations obtained by two alternatives (a and b) in terms of a particular criterion, in to a preference degree ranging from 0 to

$$p_j(a,b) = g_j\left[f_j(a) - f_j(b)\right]$$
$$0 \le p_j(a,b) \le 1$$

Be the preference function associated to the criterion, $f_j(.)$ where g_j is a non-decreasing function of the observed deviation (d) between $f_j(a)$ and $f_j(b)$. In order to facilitate the selection of a specific preference function, six basic types have been proposed (Brans 1982; Brans et al. 1985; Brans and Mareschal 1994b).

Such a generalized criterion has to be defined for each criterion. In order to facilitate the identification six types of particular preference functions have been proposed (Table 4.1).

In each case 0, 1, or 2 parameters have to be defined, their significance is clear:

q is a threshold or indifference;

p is a threshold of strict preference;

s is an intermediate value between q and p;

The indifference threshold is the largest deviation which is considered as negligible by the decision maker, while the preference threshold is the smallest deviation which is considered as sufficient to generate a full preference.

The identification of a generalized criterion is then limited to the selection of the appropriate parameters. It is an easy task (Brans and Mareschal 1994b).

Table 4.1 Types of generalized criteria (Brans and Mareschal 2005)

Types	*Generalised criterion*	*Definition*	*Parametre to fix*
Type 1: Usual Criterion	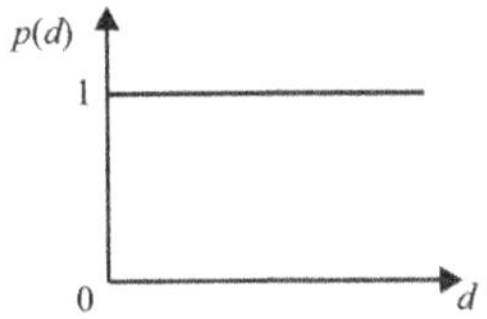	$p(d) = \begin{cases} 0, \text{ when } d \leq 0 \\ 1, \text{ when } d > 0 \end{cases}$	-
Type 2: U-shape Criterion	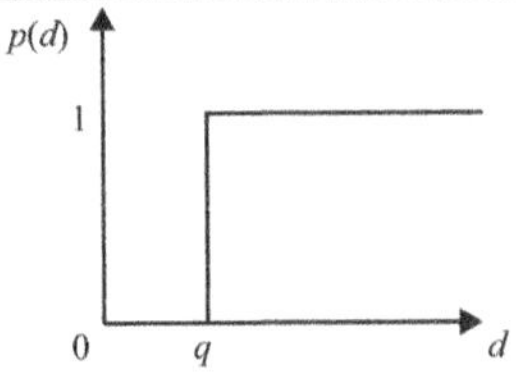	$p(d) = \begin{cases} 0, \text{ when } d \leq q \\ 1, \text{ when } d > q \end{cases}$	q
Type 3: V-shape Criterion	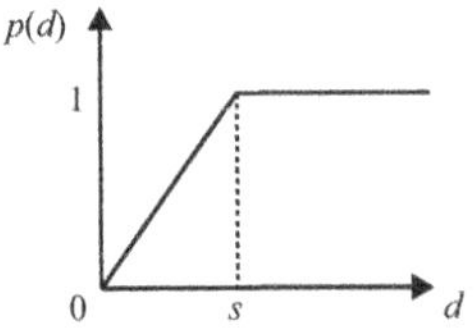	$p(d) = \begin{cases} 0, \text{ when } d \leq 0 \\ \frac{d}{s}, \text{ when } 0 < d \leq s \\ 1, \text{ when } d > s \end{cases}$	p
Type 4: Level Criterion	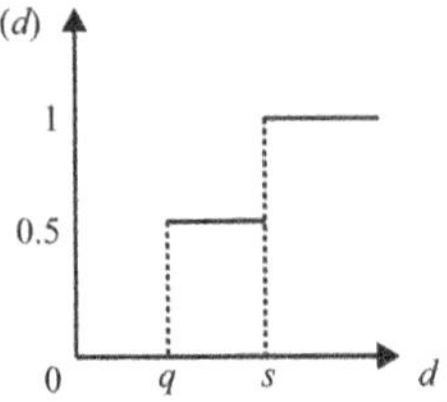	$p(d) = \begin{cases} 0, \text{ when } d \leq q \\ 0.5, \text{ when } q < d \leq s \\ 1, \text{ when } d > s \end{cases}$	$p.q$
Type 5: V-shape with indifference Criterion	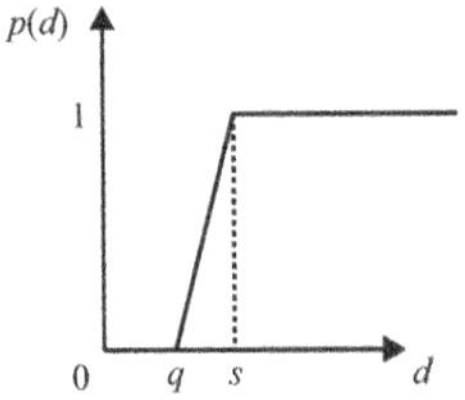	$p(d) = \begin{cases} 0, \text{ when } d \leq q \\ \frac{d-q}{s-q}, \text{ when } q < d \leq s \\ 1, \text{ when } d > s \end{cases}$	$p.q$
Type 6: Gaussian Criterion	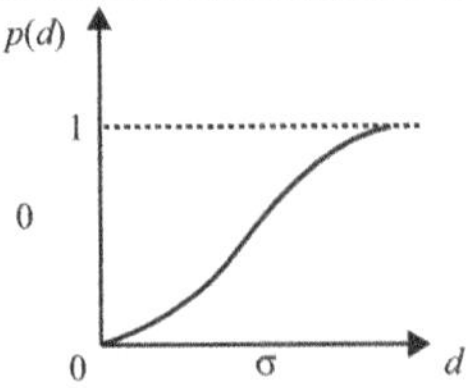	$p(d) = \begin{cases} 0, \text{ when } d \leq 0 \\ 1 - \exp(-\frac{d^2}{2\sigma^2}), \text{when } d > 0 \end{cases}$	σ

– Individual stakeholder group analysis: PROMETHEE permits the computation of the following quantities for each stakeholder r ($r = 1; \ldots; R$) and alternatives a and b:

$$\{\pi_r(a,b) = \sum_{j=1}^{k} p_j(a,b) w_{r,j};$$

$$\{\phi_r^+(a) = \sum_{x \in A}^{k} \pi_r(x,a);$$

$$\{\phi_r^-(a) = \sum_{x \in A}^{k} \pi_r(a,x);$$

$$\{\phi_r(a) = \phi_r^+(a) - \phi_r^-(a).$$

For each alternative a, belonging to the set A of alternatives, $\pi(a,b)$ is an overall preference index of a over b, taking into account all the criteria, $\phi_r^+(a)$ and $\phi_r^+(b)$. These measure, respectively, the strength and the weakness of a the other alternatives. $\phi_r^+(a)$ represents a value function, whereby a higher value reflects a higher attractiveness of alternative a. We call $\phi_r^+(a)$ the net flow of alternative a for stakeholder k.

For each stakeholder the three main PROMETHEE tools can be used to analyze the evaluation problem:

- The PROMETHEE I partial ranking: provides a ranking of alternatives. In some cases, this ranking may be incomplete. This means that some alternatives cannot be compared and, therefore, cannot be included in a complete ranking. This occurs when the first alternative obtains high scores on particular criteria for which the second alternative obtains low scores and the opposite occurs for other criteria. The use of PROMETHEE I then suggests that the decision maker should engage in additional evaluation efforts.
- The PROMETHEE II complete ranking: PROMETHEE II provides a complete ranking of the alternatives from the best to the worst one. Here, the net flow is used to rank the alternatives (Goumas and Lygerou 2000).
- The GAIA plane: The geometrical analysis for interactive aid (GAIA) plane displays graphically the relative position of the alternatives in terms of contributions to the various criteria.

The information included in matrix M is more extensive than the one in the evaluation Table 4.2, because the degrees of preference given by the generalized criteria are taken into account in M. Moreover the $g_j(a_i)$ are expressed on their own scale, while the $\phi_j(a_i)$ are dimensionless. In addition, let us observe that M is not depending on the weights of the criteria.

As the number of criteria is usually larger than two, it is impossible to obtain a clear view of the relative position of the points with regard to the criteria. We, therefore, project the information included in the k- dimensional space on a plane (Albadvi et al. 2007). Let us project not only the points representing the alternatives

Table 4.2 Single criterion nets flows

	$\phi_1(\cdot)$	$\phi_2(\cdot)$	$\dots$	$\phi_j(\cdot)$	$\dots$	$\phi_k(\cdot)$
a_1	$\phi_1(a_1)$	$\phi_2(a_1)$	$\dots$	$\phi_j(a_1)$	$\dots$	$\phi_k(a_1)$
a_2	$\phi_1(a_2)$	$\phi_2(a_2)$	$\dots$	$\phi_j(a_2)$	$\dots$	$\phi_k(a_2)$
$\vdots$	$\vdots$	$\vdots$	$\ddots$	$\vdots$	$\ddots$	$\vdots$
a_i	$\phi_1(a_i)$	$\phi_2(a_i)$	$\dots$	$\phi_j(a_i)$	$\dots$	$\phi_k(a_i)$
$\vdots$	$\vdots$	$\vdots$	$\ddots$	$\vdots$	$\ddots$	$\vdots$
a_n	$\phi_1(a_n)$	$\phi_2(a_n)$	$\dots$	$\phi_j(a_n)$	$\dots$	$\phi_k(a_n)$

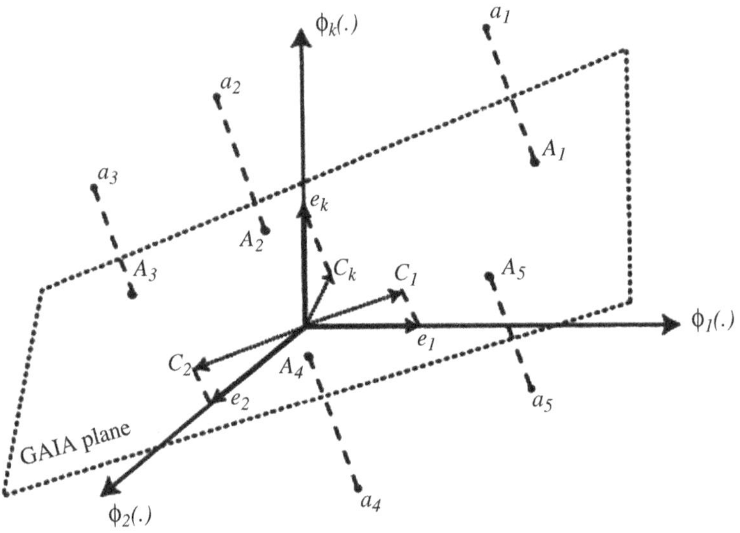

Fig. 4.1 Projection on the GAIGA plane

but also the unit vectors of the coordinate-axes representing the criteria. We then obtain (Fig. 4.1):

The GAIA plane is the plane for which as much information as possible is preserved after projection. According to the principal components analysis technique it is defined by the two eigenvectors corresponding to the two largest eigen values of the covariance matrix $M'M$ of the single criterion net flows.

Of course some information gets lost after projection. The GAIA plane is a meta model (a model of a model). Let be the quantity of information preserved.

4.4 Classification Systems of Control Aircraft Flight in Airport Roissy Charles de Gaulle

Air transport plays a major role in economic development currently facing the globalized world. In the coming decades, global air traffic, if accompanying the trend could double. The current structures of operation and air traffic control are not able to cope with this growth. Capacity limits, both for air traffic flow for that airport operation are already almost achieved in Europe, the United States, and Japan. Many airports around the world are on the verge of saturation and the slightest incident can result in a highly disadvantageous situation for all actors of air transport. Airport Paris Roissy Charles de Gaulle is a good example of platform operating at the limit of its capacity. Since its inception in 1974 until today the traffic has been multiplied by 13 and growth is currently estimated at 5% per year. This saturation to come, if not avoided, will have significant impacts on safety, efficiency, punctuality, and fluidity of air transport operations and will have significant economic consequences.

4.4.1 Different Types of Flight Control Systems of an Airplane

The advancement of research in the field of aircraft will lead to a new way of developing aircraft but also of carrying out maintenance and the emergence of new on-board services. Taking the example of an airport that seeks to classify between 4 types of control systems for piloting an airplane into 13 evaluation criteria to select the best and most suitable for its configuration using the help method to the PROMETHEE II multi-criteria decision.

- *Gander Automated Control Air Traffic System (GAATS)*: helps manage traffic, flight plans and increases situational awareness for controllers by providing the following display features such as the situation display integrating radar data and the flight plan with zoom and pan capabilities.
- *Sequencing System (SASS)*: It helps controllers allocate air traffic and landing slots to reduce delays.
- *ASDE* (Radar monitoring of surface movement): is built into the system for monitoring runway incursions and conflict alert that can be adapted to provide alerts for different situations.
- *MDS/MLAT (Display System multilateration):* is a high-performance and profitable source of positional data and identification of aircraft and vehicles equipped with transponders (NAV CANADA 2006).

4.4.2 PROMETHEE Method Analysis

- Effective criteria

The different evaluation criteria identified for the classification of flight control systems of an aircraft are in Table 4.3:

- Preference level

On base relations of upgrade of the PROMETHEE II methods, the preference of the deciders follows of the max level of satisfaction of flight control systems of an airplane choose indicated by the decider in Fig. 4.2:

Then, The GAIA analysis is obtained by projection of this information such that as few information as possible get lost. Alternatives are represented by points and, criteria by axes. The conflicting character of the criteria appears clearly in Fig. 4.3:

This figure shows the direction of the decision axis (pi) of flight control systems of an airplane.

Clusters of similar alternatives can easily be detected due to analysis of the GAIA plane (Brans and Mareschal 1994b). Criteria expressing similar preferences on the data are oriented in the same direction; conflicting criteria are pointing in opposite directions. In this case we observe, for instance, that the coupling phenomena between two parallel runways are in strong conflict with the presence of a number of runways and geometry and dimensions of the airport. It is also possible to appreciate clearly the quality of the alternatives with respect to the different criteria. SAAS is particularly good on ASDE, GAATS, and MDS, then, the ASDE and MDS are goods of GAATS.

Then, PROMETHEE II provides a complete ranking represented in the (Fig. 4.4). It is based on the balance of the two preference flows. The information looks stronger but some part of it gets lost in the process. Both PROMETHEE II help the decision maker to finalize the selection of a best compromise. A clear view of the outranking relations between the alternatives is obtained following:

In order to refer of Fig. 4.4, and we based of the criteria preference indicate of deciders, the important flight control systems of an airplane classified prior is MDS, after GAATS, after then SAAS and the final ASDE.

Table 4.3 Effective criteria

Effective criteria	Type	Weight	Unit	Preference function
Number of runways (C1)	Max	6.00	Number	U-shape, q = 2
Geometry and dimensions of the airport (C2)	Max	6.24	Scale	Usual
The relative disposition between the tracks (C3)	Max	5.49	Scale	Usual
Level and changes in flight (C4)	Min	5.18	Scale[a]	Usual
Flight planning (C5)	Max	4.34	Scale	Usual
The capacity of roads (C6)	Max	5.20	Scale	Usual
Coupling phenomena between two parallel runways (C7)	Max	4.59	Scale	Usual

Scale[a]; very low, low, average, high, very high

Preference Flows

	Φ+	Φ−	Φ
GAATS	0.4286	0.5238	-0.0952
SASS	0.3810	0.5714	-0.1905
ASDE	0.1429	0.8095	-0.6667
MDS	0.9524	0.0000	0.9524

Fig. 4.2 Preference level

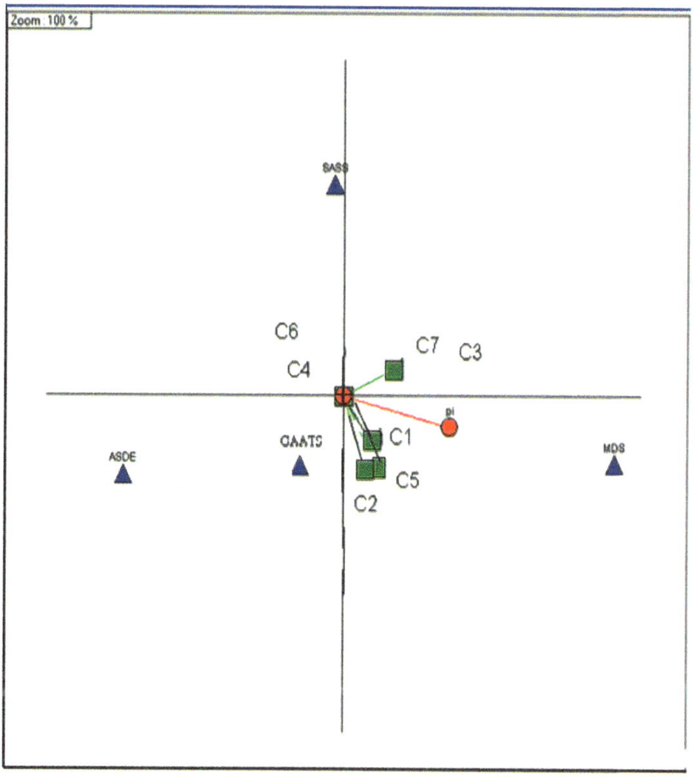

Fig. 4.3 GAIGA plane

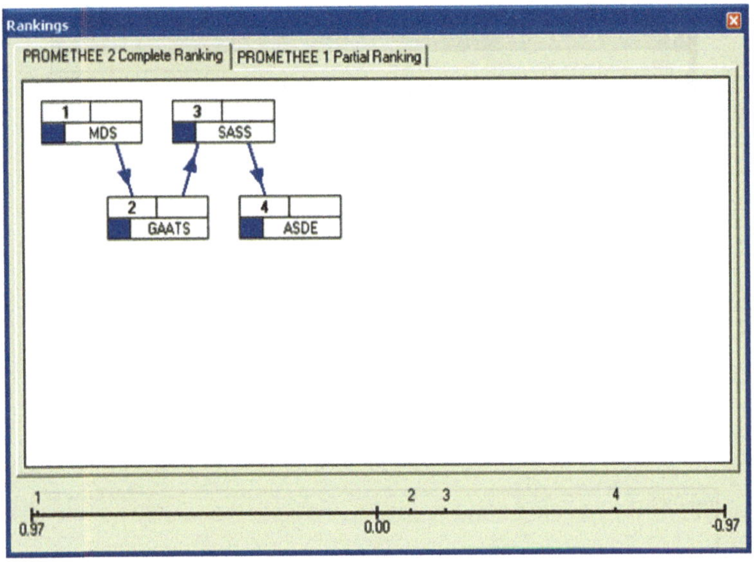

Fig. 4.4 PROMETHEE II ranking

4.5 Conclusion

This paper has argued that operational synergies can be achieved by integrating into PROMETHEE; In addition, a new weighing approach was suggested that could greatly benefit PROMETHEE. Until now, this methods tool did not provide any formal guidelines for weighing.

Then, a decision-making model is provided for selecting flight control systems of an airplane with application PROMETHEE decision-making method has been applied. The required information for implementing of this method has been gathered and analyzed through the deciders. This model has been the following results in application this model have been achieved:

- The result of the application of this model is largely dependent up on the deciders for determining the preference function of each criterion.
- By applying the PROMETHEE method for sensitivity analysis of the results, it can also help the deciders selecting flight control systems of an airplane.

References

Albadvi, A., Chaharsooghi, S. K., & Esfahanipour, A. (2007). Discrete optimization decision making in stock trading: An application of PROMETHEE. *European Journal of Operational Research, 177*, 673–683.

Al-Kloub, B., & Abu-Taleb, M. F. (1998). Application of multicriteria decision aid to rank the Jordan-Yarmouk basin co-riparians according to the Helsinki and ILC rules. *Water International, 23*(3), 164–173.

Al-Shemmeri, T., Al-Kloub, B., & Pearman, A. (1997). Model choice in multicriteria decision aid. *European Journal of Operational Research, 97*(3), 550–560.

Bana e Costa, C. A. (1990). *Readings in multiple criteria decision aid*. Berlin: Springer.

Brans, J. P., & Mareschal, B. (1994a). The PROMETHEE GAIA decision support system for multicriteria investigations. *Investigation Operative, 4*(2), 107–117.

Brans, J.-P., Mareschal, B. (2005). *PROMETHEE methods, in 'Multiple Criteria Decision Analysis: State of the Art Surveys'*. Edited by J. Figueira, S. Greco, M. Ehrgott. Springer, Chapter 5: 163–195.

Brans, J. P., & Mareschal, B. (1994b). The PROMCALC and GAIA decision support for multicriteria decision aid. *Decision Support System, 12*(1994), 297–310.

Brans, J. P. (1982). L'ingénierie de la décision. Elaboration d'instruments d'aide a la décision. Méthode PROMETHEE. In R. Nadeau & M. Landry (Eds.), *Laide a la Décision: Nature, Instrument set Perspectives D'avenir* (pp. 183–214). Quebec: Presses de Université Laval.

Brans, J. P., & Mareschal, B. (1992). PROMETHEE V–MCDM problems with segmentation constraints. *Infor, 30*(2), 85–96.

Brans, J. P., & Mareschal, B. (1995). The PROMETHEE VI procedure. How to differentiate hard from soft multicriteria problems. *Journal of Decision Systems, 4*, 213–223.

Brans, J. P., Vincke, Ph. (1985). A Preference Ranking Organisation Method: (The PROMETHEE Method for Multiple Criteria Decision-Making) Management science, *31*(6), 647–656.

Brans, J. P., Vincke, P., & Mareschal, B. (1986). How to select and how to rank projects: The PROMETHEE method. *European Journal of Operational Research, 24*(2), 228–238.

Bouyssou, D., & Perny, P. (1992). Ranking methods for valued preference relations – A characterization of a method based on leaving and entering flows. *European Journal of Operational Research, 61*(1–2), 186–194.

Chareonsuk, C., Nagarur, N., & Tabucanon, M. T. (1997). A multicriteria approach to the selection of preventive maintenance intervals. *International Journal of Production Economics, 49*(1), 55–64.

Chou, T. Y., Lin, W. T., Lin, C. Y., Chou, W. C., & Huang, P. H. (2004). Application of the PROMETHEE technique to determine depression outlet location and flow direction in DEM. *Journal of Hydrology, 287*, 49–61.

Doumpos, M., & Zopounidis, C. (2004). A multicriteria classification approach based on pairwise comparisons. *European Journal of Operational Research, 158*, 378–389.

Elevli, B., & Demirci, A. (2004). Multicriteria choice of ore transport system for an underground mine: Application of PROMETHEE methods. *Journal of the South African Institute of Mining and Metallurgy, 104*(5), 251–256.

Figueira, J. J., De Smet, Y., & Brans, J. P. (2004). MCDA methods for sorting and clustering problems: Promethee TRI and Promethee CLUSTER, Université Libre de Bruxelles. *Service deMathématiques de la Gestion, Working* Paper 2004/02.

Goumas, M., & Lygerou, V. (2000). An extension of the PROMETHEE method for decision making in fuzzy environment: Ranking of alternative energy exploitation projects. *European Journal of Operational Research, 123*(3), 606–613.

NAV CANADA. (2006). Techwatch bulletin nav canade, Numéro 20, Septembre 2006.

Huylenbroeck, G. (1995). The conflict-analysis method – Bridging the gap between ELECTRE, PROMETHEE and ORESTE. *European Journal of Operational Research, 82*(3), 490–502.

Yan, J., Dagang, T., & Yue, P. (2007). Ranking environmental projects model based on multicriteria decision-making and the weight sensitivity analysis. *Journal of Systems Engineering and Electronics, 18*(3), 534–539.

Yang, J. B., & Sen, P. (1997). Multiple attribute design evaluation of complex engineering products using the evidential reasoning approach. *Journal of Engineering Design, 8*(3), 211–230.

Yang, J. B., & Xu, D. L. (2002). On the evidential reasoning algorithm for multiple attribute decision analysis under uncertainty. *IEEE Transactions on Systems Man and Cybernetics Part A – Systems and Humans, 32*(3), 289.

Vincke, J. P., & Brans, P. (1985). A preference ranking organization method. The PROMETHEE method for MCDM. *Management Science, 31*, 641–656.

Wiecek, M. M., Ehrgott, M., Fadel, G., & Figueira, J. R. (2008). Multiple criteria decision making for engineering. *Omega, 36*, 337–339.

Chapter 5
The Multi-Criteria Approach Assessment Human Risks: Application the Analytical Hierarchy Process and PROMETHEE Methods

5.1 Introduction

The actual systems are more and more complex, because they integrate a variety of technologies. The system need often long period of development. When we take a change on requirements, the objectives are to improve the functionality, the cost, or the delay of systems; but unfortunately, this modification can affect a problem with others requirements those involved with the safety of the system.

The requirements change occurs in two main cases. The first concerns revising/updating existing requirements that led to an actual version of the systems to adapt to new environment (Larsen and Buede 2002).

The second is when new technology is being developed and new requirements are implemented consequently for reasons of cost or feasibility.

Then, the change of requirements has the effect on the change of the level of the risk in the system. Then the change in requirements has the effect of changing the level of risk in the system. These risks cover accidents of a technological nature, which evoke a system of rules allowing technical problems to be noted. Then, this technical risk analysis gave rise to several failures (Leena and Paula 2004). Indeed, different generated sucking accidents complex systems are mostly the issue of a chain of events of which each link taken isolation appears as minor, and of which some links are human errors.

Then, the proposed document makes it possible to contribute to improving safety with the assessment of human risk by applying a multi-criteria decision support approach.

L. Azzabi et al., *The Multi-Criteria Approach for Decision Support*, International Series in Operations Research & Management Science 300, https://doi.org/10.1007/978-3-030-57262-4_5

5.2 Human Risks

The concept of human error is very elusive. At a closer look, the frequent allocation of accidental causes to human error appears to be subjective and guided by the tool box of the analyst. This is a simple reflection of the nature of causal analysis and the fact that no objective stop rule exists to terminate the causal back tracking in search of a root cause. The search stops when an event is found for which a cure is known to the analyst. Classic human factors evaluation involves task analysis and experimentation to better understand the effect of the task design on human performance. Only recently tools have been developed to support human factors analysis; when the need for automated prediction of human performance prior to system development became apparent. At the heart of human factors analysis resides human reliability. Recent accident statistics show that 80% of failures are attributed to human error (Gregoriades and Sutcliffe 2008).

A deeper analysis of accident causation indicates that the observed coincidence of multiple errors cannot be explained by a stochastic coincidence of independent events. Accidents are more likely caused by a systematic migration toward accident by an organization operating in an aggressive, competitive environment. Commercial success depends on exploitation of the benefit from operating at the fringes of the usual, accepted practice (Rasmussen 1999).

The importance of the human factors contribution to safety has been demonstrated over the past decades by the often-quoted examples of the contribution of human failure to major accidents within the process industry. Given this demonstration, it is surprising that the value of assessing human factors has not yet been comprehensively accepted throughout the industry. So what is a human factor? (Hughes and Weichel 2004). A recent definition of human factors as related to the process industry and safety executive:

"Human factors refer to environmental, organizational and job factors and human and individual characteristics which influence behavior at work in a way which can affect health and safety" (Health and Safety Executive 1999).

The use of functional analysis methods was a first tentative answering of the risk (Fadier and Ciccotelli 1999), when carrying out a functional analysis; the attention is essentially focused on the technical functions. The operator is not taken in account or just summarily (ergonomics of the handful, colors used in the interface. . .), and he (or she) often inherits functions that the technical system cannot fill.

However, the new regulations concerning work safety require that the conception integrates the fact that the work equipment can be badly used, subject to breakdown, or dangerous for the operator. The instantiation of the European Directives and standards is supposed to suppress a majority of these risks (Neboit 2003). Safe design in this context also means a design that allows and conditions, as far as feasible, safe use across the whole life cycle of the product, from manufacture, construction, transportation, and installation, through use, maintenance, to decommissioning, demolition, and disposal. While the case for incorporating safety at the design stage is strong, it is not universally accepted, also not by all suppliers, who may seek to limit their liability by pushing the main decisions about safety over to the user (Hale et al. 2003).

Then, in specific conditions, these operating modes appear as palliative activities to maintain the performance of the system, sometimes to the detriment of the health and safety of the operators as their primary objective is the optimization of the system. In addition, they are the result of a compromise faced with managing various system constraints and "operating deviations" in the process characterizing a migration towards tolerance limit thresholds from the performance and safety points of view (Fadier et al. 2003). They are constructed and adapted by the organization using the system (evolution of the Human-Task system) and evolve with the environment and with time. They, therefore, have a dynamic character that, depending on the situation, can tend towards a reduction in uncertainty by increasing the room for maneuver of the operators or, they may lead to increase in uncertainty by reducing the room for maneuver of the operators. By uncertainty here we mean the possibility of controlling the situation (anticipation, diagnosis, actions, dynamic aspect of the situation) or of losing control of it. Uncertainty evolves in any given situation and is partly linked to the room for maneuver of the operators. Indeed, the latter refers to the span of initiative and the span of tolerance that operators actually have to regulate the operation of the Human–Machine system (Ayadi et al. 2008). Room for maneuver depends on the rules, the instructions and means given to the operators, their skills, and the effective characteristics of the situation. Analysis of the regulation of the process of social action shows the difference between two possibilities of room for maneuver: autonomy and discretion.

5.3 What Is Risk Assessment?

Risk assessments, whether they pertain to information security or other types of risk, are a means of providing decision makers with information needed to understand factors that can negatively influence operations and outcomes and make informed judgments concerning the extent of actions needed to reduce risk (Ayadi et al. 2008).

A risk assessment is an important step in protecting your workers and your business, as well as complying with the law. It helps you focus on the risks that really matter in your workplace—the ones with the potential to cause real harm. In many instances, straightforward measures can readily control risks (Health and Safety Executive 2006).

Risk assessment can be qualitative or quantitative. In the presence of known hazards, quantitative assessments can be done. But in many cases, quantitative data will be incomplete or even absent. Types, subtypes, and variants of infectious agents involving different or unusual vectors, the difficulty of assays to measure an agent's amplification potential, and the unique considerations of genetic recombinants are but a few of the challenges to the safe conduct of laboratory work. In the face of such complexity, meaningful quantitative sampling methods are frequently unavailable. Therefore, the process of doing a risk assessment for work with biohazardous materials cannot depend on a prescribed algorithm (Haimes 1996).

Many industries do risk assessments. Depending on the circumstances, a risk to creative intuition is a careful analysis of possible systemic weaknesses. The assessment can be retrospective or prospective, quantitative or qualitative, or a combination. A quantitative risk assessment, using valid data, can be more objective and more useful over time than a qualitative risk assessment developed from assumptions and random case studies. Retrospective risk assessments have the benefit of drawing on data from past events to help anticipate future problems. Although the past is not always a reliable indicator of the future, consistent patterns can emerge in data and crime data is no exception. Prospective risk assessments attempt to see into the future without the benefit of historical data. The risk assessment process when little or no data is available involves using whatever information is known to anticipate real or potential outcomes. This approach relies primarily on qualitative rather than quantitative indicators. The best risk assessment methodology uses a combination of approaches in order to capture all that is known and as much as possible about what is not known. There are likely to be illicit financing methods being used that have not been detected by financial institutions or law enforcement, and so will not show up in the data gathered from criminal investigations or financial institution currency or transaction reporting. And there may be other illicit financing options that even the criminals have not yet discovered. In the absence of data or case studies identifying these methods, financial institutions and competent authorities must rely (Pate-Cornell 1998).

5.4 The Process of Risk Assessment?

In many organizations, the risks are well known and control measures are necessary. The risk assessment process is identified in five steps (Fig. 5.1):

5.4.1 Look for the Hazards

Hazard identification, this first step in risk assessment, consists in collecting data from different sources to determine whether a substance is toxic. It involves gathering and examining data from toxicological and epidemiological studies (Ingram 2003).

Then, the Hazard identification indicates whether exposure to a substance causes a harmful health or environmental effect and the nature of the effect. Hazardous substances are identified by analyzing the wastes that will be fed into the incinerators to determine what kind of air pollutants might be produced during the incineration process, and by collecting emissions information during trial burns.

Fig. 5.1 Process risk assesment

After to have to collect all indispensable information of hazard, it is to schedule that by major order using the multi-criteria approach to help the decision such as the PROMETHEE method. A multiple-criteria decision problem is one that, having a defined set of consequences (alternatives), A, a family of criteria depending on A, designated as C, and a set of preference relations G, intends to find a subset of A containing the best consequences, to assign the alternatives into predefined categories or rank them (Roy 2000). Each of these objectives defines a different multiple-criteria problem:

(a) Selection; (b) classification or sorting; (c) ranking. The PROMETHEE method (preference ranking organization method for enrichment evaluation) is a multi-criteria decision-making method developed by Brans (Sylvain et al. 2003). It is a quite simple ranking method in conception and application compared with other methods for multi-criteria analysis. It is well adapted to problems where a finite number of alternatives are to be ranked considering several, sometimes conflicting criteria. The implementation of PROMETHEE requires two additional types of information, namely:

- Information on the relative importance of the criteria considered,
- Information on the decision-makers preference function, which he/she uses when comparing the contribution of the alternatives in terms of each separate criterion.

The following steps are required for the implementation of the method:

1. Alternatives are compared in pairs for each criterion. The preference is expressed by a number in the interval [0, 1] (0 for no preference or indifference to, 1 for strict preference). The function relating the difference in performance to preference is called the generalized criterion and it is determined by the decision maker.
2. A multi-criteria preference index is formed for each pair of alternatives as a weighted average of the corresponding preferences computed for each criterion. The index Π (a,b) (in the interval [0, 1]) expresses the preference of alternative a over b considering all criteria. The weighting factors express the relative importance of each criterion and are chosen by the decision maker.
3. Alternatives can be ranked according to:
    - The sum of indices π (a, i) indicating the preference of alternative a over all the others. It is termed "leaving flow" ∅ +(a) and shows how "good" is alternative a.
    - The sum of indices π (i, a) indicating the preference of all other alternatives compared to a. It is termed "entering flow" ∅- (a) and shows how "inferior" is alternative a.

Then, the method PROMETHEE II extends this classical approach by modeling multi-criteria decision the preferences through a preference function *pi*(*di*) for the *i* th criterion, in such way that it reflects the preference level of a over b, from 0 to 1. If *pi*(*di*) = 0, both alternatives are considered indifferent to each other. If *pi*(*di*) = 1, a is strictly preferred to b. These functions can have an indifference interval, limited by a threshold ti, specified by multi-criteria decision which allows him to be indifferent between two alternatives not only when they have the same consequences considering the *i* th objective function (*di* = 0), but also when they are just similar to each other (0 < *di* < *ti*). Furthermore, these functions can have a smooth transition between indifference and strict preference. Values between 0 and 1 express the preference intensity of the multi-criteria decision judgment, in such way that *pi*(*di*)~0 indicates weak preference and *pi*(*di*)~1 indicates strong preference. Brans et al. (1986) proposes four main types of preference functions (it presents two other types that can be derived from these ones), which cover most of the practical situations.

PROMETHEE requests additional information. For each criterion a specific preference function must be defined. This function is used to compute the degree of preference associated to the best action in case of pair-wise comparisons (Pierre and Bertrant 1986).

As in PROMETHEE different types of preference functions can be used depending on the nature of the data. Because they are somewhat different from the one used in PROMETHEE they are represented as type A to F. This method includes tools that do not have a very good axiomatic basis but show a pragmatic realistic quality in regard to the decision settings often encountered, many new concepts are found in this method such as decision problematic, discrimination power of a criterion (Marc 1999). The criteria and their preference function are represented in Fig. 5.2:

PROMETHEE methods are characterized by an infinite discriminating power. Any difference in score implies immediately a total preference or in other words the preference indicator $P\ (a,b)$ will only depend upon the sum of weights of those criteria for which a is better than b. The same holds for the second type of preference function, the quasi-criterion or U-shaped criterion, but here an indifference threshold q, allowing a margin of error in the evaluation scores, is considered (Colson 2000).

Criterion type	*Preference function*	*Parametres to be defined*
Type A : 0-1 criterion $\hat{e}\,(a,b) = \begin{cases} 0 \ if \ \lvert e(a) - e(b)\rvert = 0 \\ 1 \ if \ \lvert e(a) - e(b)\rvert > 0 \end{cases}$	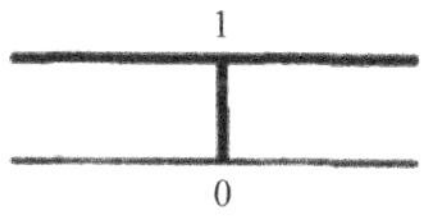	
Type B : 0-1 criterion with indefference ares $\hat{e}\,(a,b) = \begin{cases} 0 \ if \ \lvert e(a) - e(b)\rvert = q \\ 1 \ if \ \lvert e(a) - e(b)\rvert > q \end{cases}$	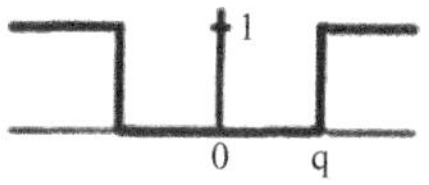	- Indifference threshold q
Type C : Multilevel criterion $\hat{e}\,(a,b) = \begin{cases} 0 \ if \ \lvert e(a) - e(b)\rvert \leq q \\ \frac{k}{n} \ if \ k1 < \lvert e(a) - e(b)\rvert \leq k2 \\ 1 \ if \ \lvert e(a) - e(b)\rvert > p \end{cases}$	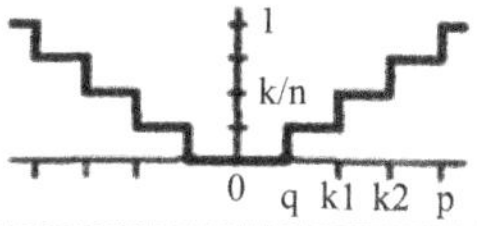	- Indifference threshold **q** - Interval thresholds **k** - Preference threshold **p**
Type D : Linear criterion $\hat{e}\,(a,b) = \begin{cases} 0 \ if \ \lvert e(a) - e(b)\rvert \leq q \\ \frac{\lvert e(a) - e(b)\rvert}{p} \ if \ q < \lvert e(a) - e(b)\rvert \leq p \\ 1 \ if \ \lvert e(a) - e(b)\rvert > p \end{cases}$	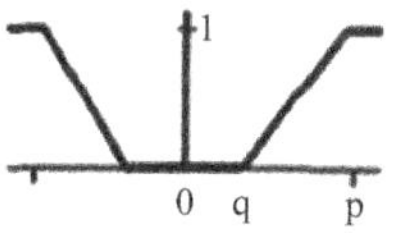	- Indifference threshold **q** - Preference threshold **p**
Type E : Rank order criterion $\hat{e}\,(a,b) = \begin{cases} 0 \ if \ \lvert e(a) - e(b)\rvert = 0 \\ \frac{k}{n} \ if \ \lvert e(a) - e(b)\rvert = k \\ 1 \ if \ \lvert e(a) - e(b)\rvert > p \end{cases}$	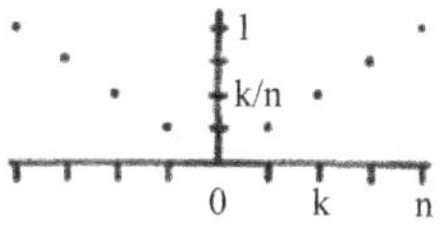	
Type F : Gaussian criterion $\hat{e}\,(a,b) = 1 - e^{-[\lvert e(a) - e(b)\rvert^2 / 2S^2]}$	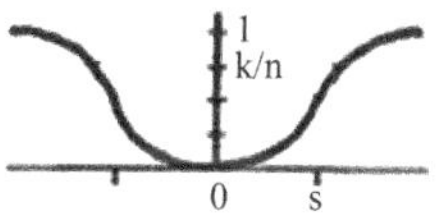	- Flexure point S

Fig. 5.2 Criteria and their preference function (Huylenbroeck 1995)

5.4.2 *Decide Who Might Be Harmed*

For each hazard, determining who might be injured will help identify the best way to manage the risk. Identifying the danger/risk relationship is obtained by applying the fault tree analysis (FTA), the objective of which is to determine the causes responsible for the appearance of human risk. Fault Tree Analysis (FTA) is one of the most widely used techniques in this context, it is intuitive for practitioners due to their hierarchical structure and the familiar logical symbols. They allow a variety of qualitative and quantitative analyses. Fault Tree Analysis (FTA) has been used for several decades in the context of mechanical or electrical systems (Bernhard et al. 2007).

FTA is a method for analyzing causes of hazards. FTA can describe the dynamic accident process of occurrence and development. It is convenient to find out the direct and indirect reasons and combinations of these reasons. Qualitative analysis can identify the priority of hidden reasons and hazards and can predict the probability of occurrence of accidents. The formulation and analysis of the fault tree require a solid mathematical foundation and considerable professional skills. ETS is a method of deductive reasoning, the logical relationship between potential accidents and the corresponding reasons can be represented by tree diagrams. By qualitatively and quantitatively analyzing FTA, we can find the main reasons, provide a reliable basis for determining the safety countermeasures, in order to predict and prevent accidents (Huacan and Menglan 2014).

The tree is written as a Boolean expression to show the specific combination of identified basic events sufficient to cause the undesired top level event. If the individual probabilities for all the basic events are known (not feasible in most abstract cases), the frequency of the top event can be calculated (Fig. 5.3).

Often the most difficult part of creating a fault tree is the determination of the top level event. The selection of the top event is crucial since hazards in the system will not be comprehensive unless the fault trees are drawn for all significant top level events.

FTA est une analyse pilotée par l'état dans laquelle les entrées d'une porte logique représentent l'état d'une pièce et / ou d'un autre facteur inclus dans l'analyse (Goetsch 1997). D'autres facteurs peuvent inclure des éléments tels que la formation, les outils, l'équipement de sécurité, la supervision, etc. La sortie d'une porte logique est un état logique qui représente une condition qui existe dans le système. Un événement se produit lorsque la sortie d'une porte change d'état (Villemeur 1992).

Fault Tree Analysis (FTA) is a widely accepted model that graphically shows how influence factors (in general component failures) contribute to some given hazard or accident. They provide logical connectives (called gates) that allow decomposing the system-level hazard recursively. The most fundamental gates are the AND gate and the OR gate. Both gates are shown, more gates have been proposed in literature, but some of them have been continuously under discussion, because they cannot be mapped correctly onto pure propositional logic or pose difficulties to some of the used evaluation algorithms (Papadopoulos and Maruhn 2001) (Table 5.1).

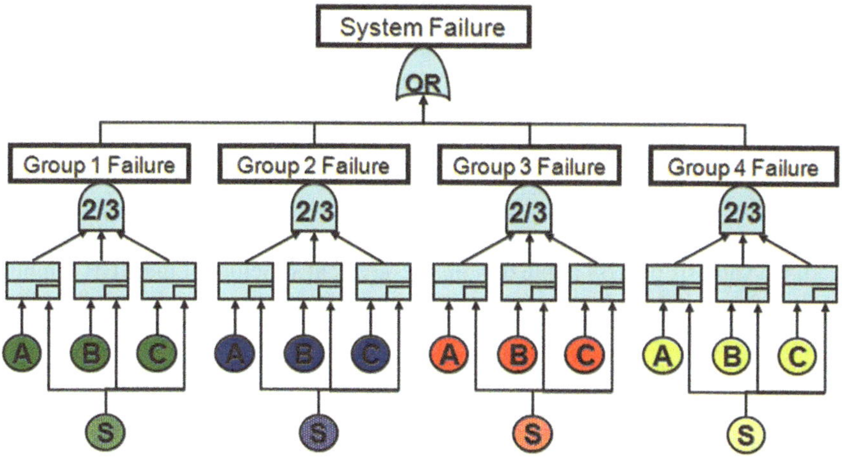

Fig. 5.3 Structure of fault tree analysis

Table 5.1 Description of FTA Gates

Description	Picture	Truth table Input A	Input B	Output
AND Gate. The AND gate indicates that the output occurs if and only if all of the input events occur		T	T	T
		T	F	F
		F	T	F
		F	F	F
OR Gate. The OR gate indicates that the output occurs if and only if at least one of the input events occur		T	T	T
		T	F	T
		F	T	T
		F	F	F

5.4.3 *Evaluate the Risk*

If the hazard identification process produces evidence of a hazard, then a hazard evaluation is performed. The purpose of this step is to calculate, if possible, the dose at which a harmful effect will occur.

This step is realized by multi-criteria methods "Analytical Hierarchy Process: AHP." Analytic Hierarchy Process is a powerful and flexible method used for making decisions that help determine the priorities, and leads to making optimal decisions in cases where aspects of quantity and quality are being taken into consideration (Saaty 2001).

Reducing complex decision-making to a comparison between alternative pairs, and synthesizing the obtained results, AHP not only helps to make decisions, but also leads to the rational decision. The Analytic Hierarchy Process for decision-making is a theory of relative measurement based on paired comparisons used to derive normalized absolute scales of numbers whose elements are then used as priorities (Saaty 2002). Matrices of pair-wise comparisons are formed either by providing judgments to estimate dominance using absolute numbers from 1 to 9 fundamental scales of the AHP, or by directly constructing the pair-wise dominance ratios using actual measurements. The AHP can be applied to both tangible and intangible criteria based on the judgments of knowledgeable and expert people, although how to get measures for intangibles is its main concern. The weighting and adding synthesis process applied in the hierarchical structure of the AHP combines multidimensional scales of measurement into a single "uni-dimensional" scale of priorities. In the end we must fit our entire world experience into our system of priorities if we are going to understand it (Saaty 2007). Steps of the method AHP are as follows:

- Construction of the hierarchy, it i s an abstraction of the structure of the used problem to study; the interaction with components of problem and their effect on the final solution, she permits to decompose the problem in a hierarchy of data inter-bound. A top of the hierarchy, one finds the objective, and in inferior level, elements contributing to fetch this objective, the last level are that of actions (Fig. 5.4).
- Carry out elementary pairwise comparisons of each hierarchical relative level to an element of the higher hierarchical level. This step makes it possible to build a matrix of comparisons, this matrix is obtained by numerical values according to

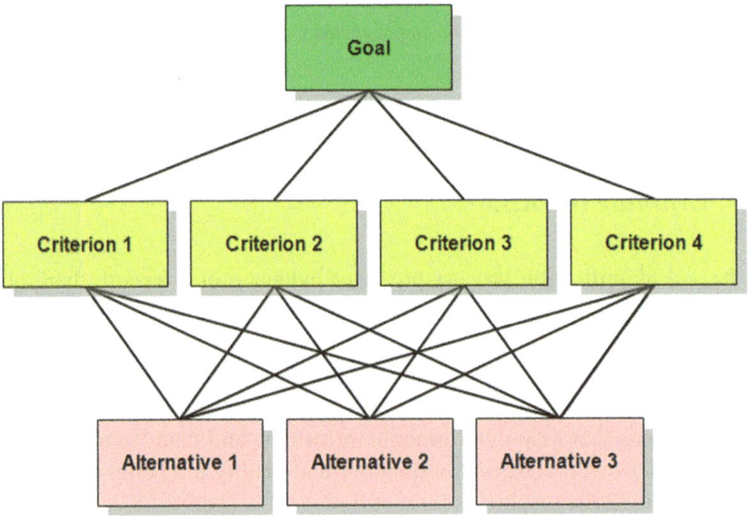

Fig. 5.4 Construction of the hierarchy

Table 5.2 Saaty scales

Importance grade	Define
1	Importance equalizes of both elements
3	Importance weak person of a relative element to another
5	Importance strong or determinant of a relative element to other
7	Importance attested of a relative element to another
9	Importance absolved of a relative element to another
2,4,6,8	Intermediate values with two values neighbor

the scale of Saaty (scale of binary comparisons), while respecting the principle of reciprocity (Table 5.2):

$$(Ea, Eb) = \frac{1}{Pc(Eb, Ea)}$$

- To determine the relative elemental importance calculating primary vectors to correspond of the maximal values of comparisons matrix.
- To verify the judgments coherence. One to calculate at first, the indicator coherence IC

$$IC = \frac{\lambda max - n}{n-1}$$

- where λ max is the primary maximal value running in matrix of comparisons by pairs and n: are a large number of comparative elements.

  Then; the ratio of coherence (RC) defines by

$$RC = 100.\frac{IC}{ACI}$$

where ACI is the means coherence indicator of obtained generating random matrix of judgment equalizes height. The means of indicator coherence is identified in the following table (Table 5.3):

A value of RC inferior to 10% is generally acceptable; otherwise, comparisons by pairs must be examined again to reduce the incoherence.

- to settle the relative performance of each action:

$$P_k\left(e_1^k\right) = \sum_{J=1}^{nk-1} P_{K-1}\left(e_i^{k-1}\right) P_k \frac{e_i^k}{e_i^{k-1}}$$

Table 5.3 Means coherence indicator

Matrix dimension	Random coherence ACI
1	0.00
2	0.00
3	0.58
4	0.90
5	1.12
6	1.24
7	1.32
8	1.41
9	1.45
10	1.49

– with: $P_k\left(e_1^k\right)=1$
and: nk – 1 are a large number of elements of the hierarchic level k——1,
Pk$\left(e_i^k\right)$ is the terms priority to the element e^i to the hierarchic level *k* (Saaty 1989).

5.4.4 *Record Your Finding*

The next step is to put in place the preventive and protective measures. It is important to involve the workers and their representatives in the process. Effective implementation involves the development of a plan specifying:

- The measures to be implemented;
- Who does what and when?
- When it is to be completed.

It is essential that any work to eliminate or prevent risks is prioritized.

The step records your finding, combines the information collected during the first three steps, and determines the likelihood that humans or animals will experience any of the health effects associated with a substance, or that the environment will be harmed.

Then, the generation of possible solutions for detected risks by the brainstorming methods.

Brainstorming is a group creativity technique designed to generate a large number of ideas for the solution to a problem (Osborn 1963).

Although brainstorming has become a popular group technique, researchers have generally failed to find evidence of its effectiveness for enhancing either quantity or quality of ideas generated. Because of such problems as distraction, social loafing, evaluation apprehension, and production blocking, brainstorming groups are little more effective than other types of groups, and they are actually less effective than individuals working independently (Nijstad et al. 2003).

5.4.5 *Review Your Assessment*

All forms of assessment should be reviewed on a regular basis. The frequency of review should be based on the hazards present. Tasks involving significant hazards should be reviewed on a much more frequent basis than lower hazard tasks. Then, Step five of the steps to risk assessment is to review the risk assessment "from time to time"; in this section the factors prompting a new risk assessment are set out (Neathey et al. 2006).

5.5 Application the Process Risk Assessment in the Industry Treatment Gas (Fig. 5.5)

The Gas decomposes features mixed with air or other substances oxidant and provides areas of flammability. Then, the structure of the natural gas market is very sensitive, provided that this area covers a significant number of accidents per year. The risks in the field Gas Processing occupy a special place in the study of security; they have largely shaped the landscape of risk in the field of security analysis, and are generally caused profound regulatory changes.

Fig. 5.5 The treatment plant gas

5.5.1 Look for the Hazards

This step permits to identify five hazards proceeding to the Gas industry, these hazards are identified by the figure as following (Table 5.4):

After that to identify the hazards, we classified these hazards by PROMETHEE, it is important to identify the list of criteria in Table 5.5):

On base relations of upgrade of the PROMETHEE II methods, the preference follow of the maximum and minimum level of the criteria indicated by the decider in Table 5.6:

Obviously the PROMETHEE II ranking is influenced by the weights allocated to the criteria. For the following weight distribution is similar (20, 20, 20, 20, and 20). Then, the ranking of the last five hazards is now completely opposite (Fig. 5.6).

The GAIA analysis is obtained by projection of this information such that as few information as possible get lost. Hazards are represented by points and, criteria by axes. The conflicting character of the criteria appears clearly (Fig. 5.7): criteria expressing similar preferences on the data are oriented in the same direction; conflicting criteria are pointing in opposite directions. In this case we observe, for instance, that the criteria C4 (Percentage of loss of life) is in strong conflict with the criteria C3 (Level of worker protection during exposure to combustible). It is also possible to appreciate clearly the quality of the alternatives with respect to the different criteria. H4 (Failure in the tank truck will lead to the loss of containing the tank) and H1 (Instability of the capacitor line connecting the receiver pellet in the area of stabilization capacitor) are particularly good on H3 (Catastrophic failure of the receiver chip) and H5 (Exposure to sulfur fuel used) look good on H2 (Lack of measurement of natural gas).

Table 5.4 Identification Hazards

Hazards	Description
H1	Instability of the capacitor line connecting the receiver pellet in the area of stabilization capacitor
H2	Lack of measurement of natural gas
H3	Catastrophic failure of the receiver chip
H4	Failure in the tank truck will lead to the loss of containing the tank
H5	Exposure to sulfur fuel used

Table 5.5 Identification of the criteria

Criteria	Description
C1	Reputation of the hazards
C2	The number of maintenance
C3	Level of worker protection during exposure to combustible
C4	Percentage of loss of life
C5	Response times for breach of gas

Table 5.6 Level of the criteria

Criteria	C1	C2	C3	C4	C5
Min/max	Min	Max	Max	Min	Max
H1	34	1	1	55	10
H2	21	5	3	66	22
H3	12	3	5	80	32
H4	15	2	5	70	15
H5	17	5	4	68	60

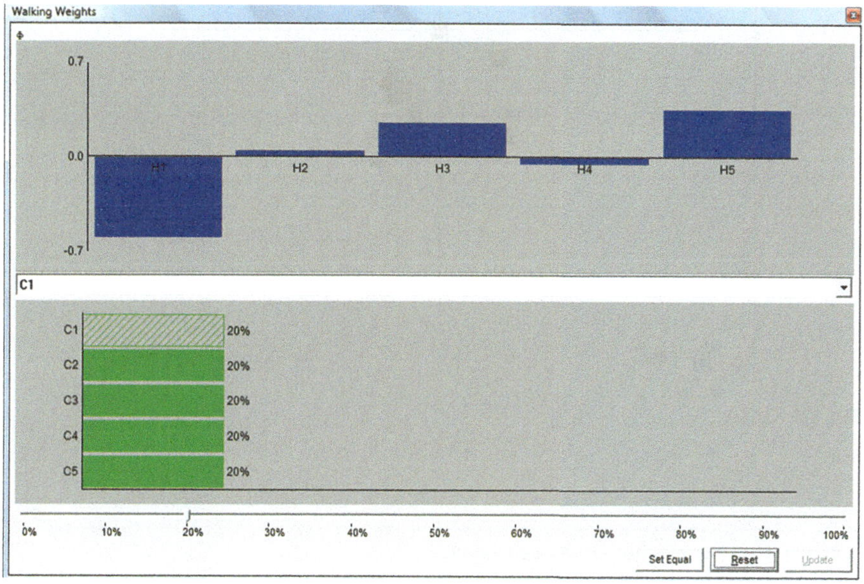

Fig. 5.6 Walking weights

Then, PROMETHEE II provides a complete ranking represented in Fig. 5.8. It is based on the balance of the two preference flows. The information looks stronger but some part of it gets lost in the process. Both PROMETHEE II help the decision maker to finalize the selection of a best compromise. A clear view of the outranking relations between the hazards is obtained following:

In order to refer to figure 8, the significant hazards are classified by the PROMETHEE method are H5 (Exposure to fuel sulfur), H3 (Catastrophic failure of the receiver chip.), H2 (Lack of measurement of natural gas), H5 (Exposure to sulfur fuel used), and at the end H1 (Instability of the capacitor line connecting the receiver pellet in the area of stabilization capacitor).

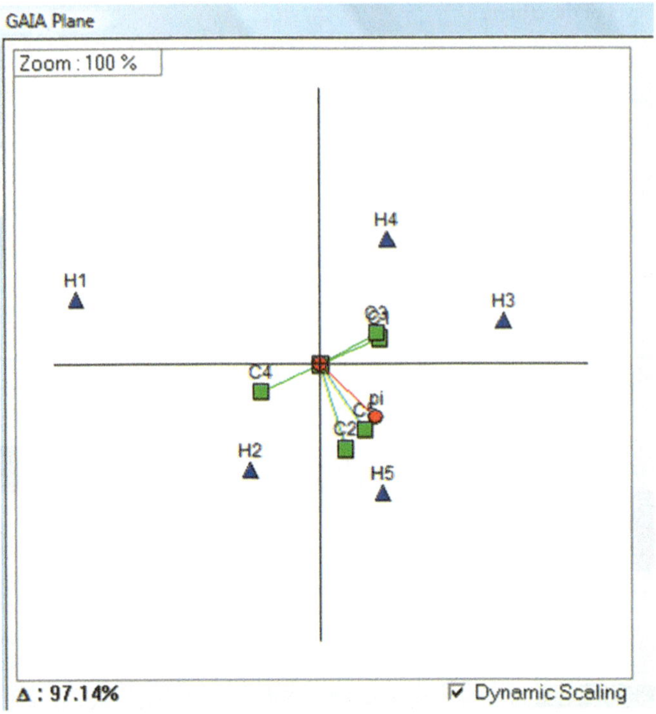

Fig. 5.7 GAIA

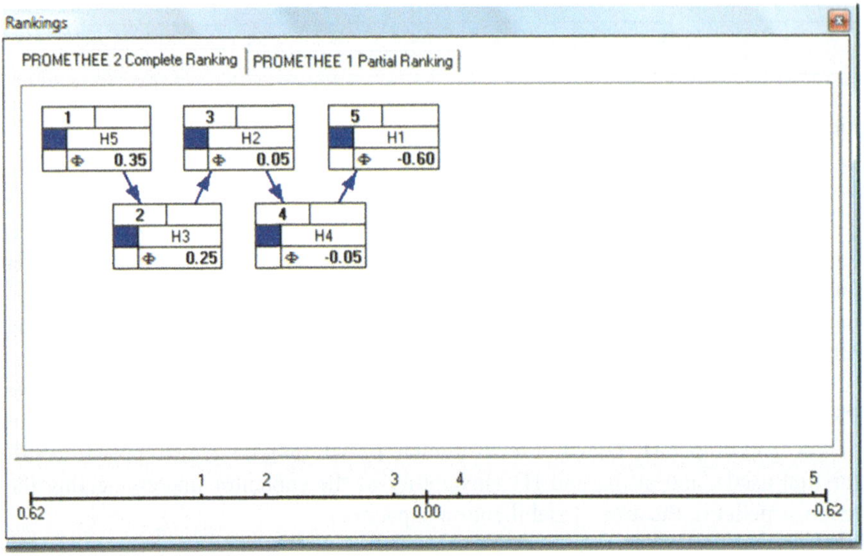

Fig. 5.8 PROMETHEE II ranking

5.5.2 Decide Who Might Be Harmed

The identification of the relation between the hazards H5: Exposure to sulfur fuel used and the risks procured by this hazard is realized by the Fault Tree Analysis (Fig. 5.9).

5.5.3 Evaluate the Risk

After classifying the different risks provided by the PROMETHEE method (H5, H3, H2, H5, and at the end H1). This step now consists of evaluating the weight of the severity of each risk on the worker by referring to the AHP method.

Then, the decomposition of the problem of classifying risks in a hierarchy inter-related elements of the hierarchical structure of the problem is as follows (Fig. 5.10):

The pair-wise comparison of elements of each hierarchical level relative to an element of higher level is as follows (Fig. 5.11):

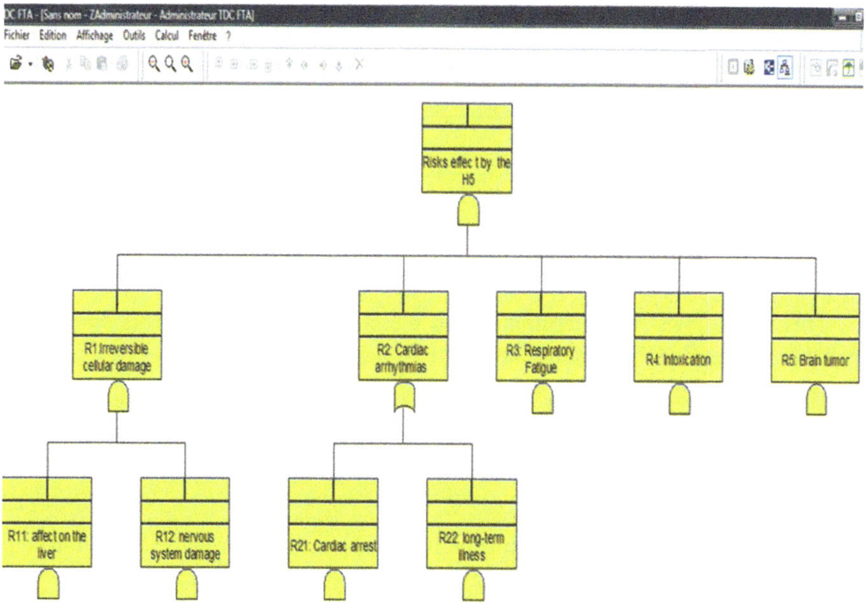

Fig. 5.9 Risks effect by H5

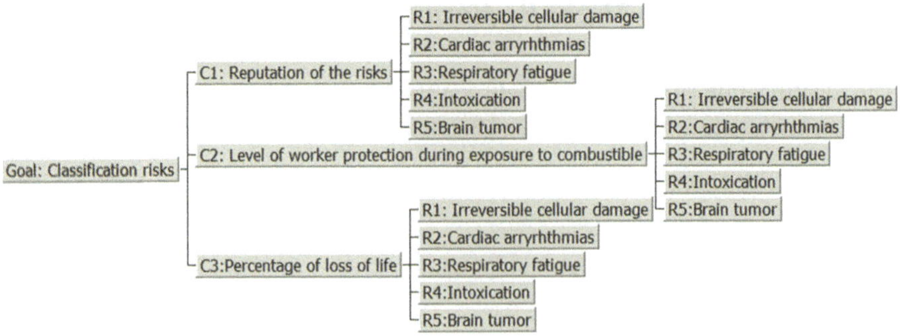

Fig. 5.10 Hierarchical structure of the problem of risk classification

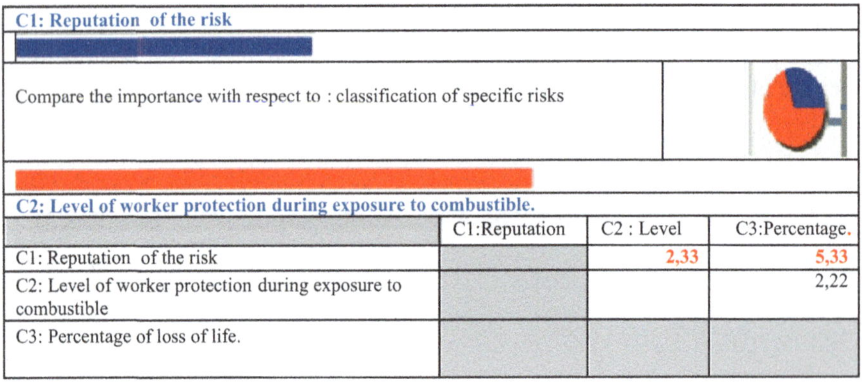

C1: Reputation of the risk

Compare the importance with respect to : classification of specific risks

C2: Level of worker protection during exposure to combustible.

	C1:Reputation	C2 : Level	C3:Percentage.
C1: Reputation of the risk		2,33	5,33
C2: Level of worker protection during exposure to combustible			2,22
C3: Percentage of loss of life.			

Fig. 5.11 Pair-wise comparison of criteria

This step allows the determination of the comparison matrix. This above, the matrix of criterion C1: Reputational risk Fig. 5.11.

- Determination of the vector for criterion C1: Reputation Risk (Fig. 5.12):
- Determination of the vector for criterion C2: Level of worker protection during exposure to combustible (Fig. 5.13):
- Determination of the vector for criterion C3: Percentage of loss of life (Fig. 5.14):

After to identify the vector corresponding of the three criteria, the following step consist to identify the performance (Fig. 5.15):

The ranking by the AHP method has established a risk classification. From this classification we consider the risk the most important is R5: Brain tumor. The final ranking is as follows (Fig. 5.16):

Then in the following step (Record your finding), we search the solution of the important and priority risks R5.

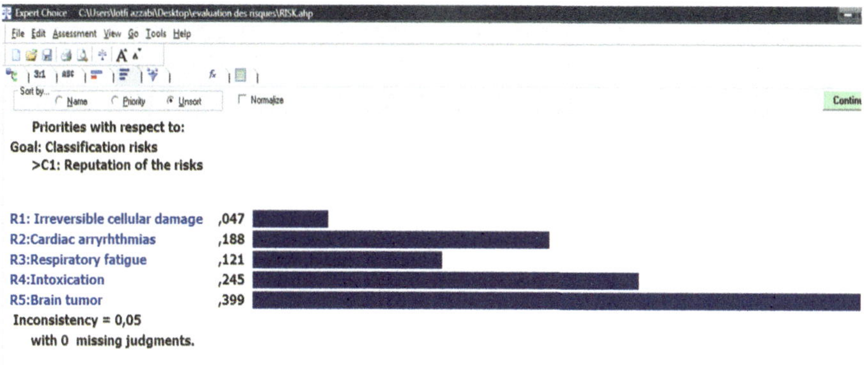

Fig. 5.12 Vector corresponding to the criterion C1 "Reputation Risk"

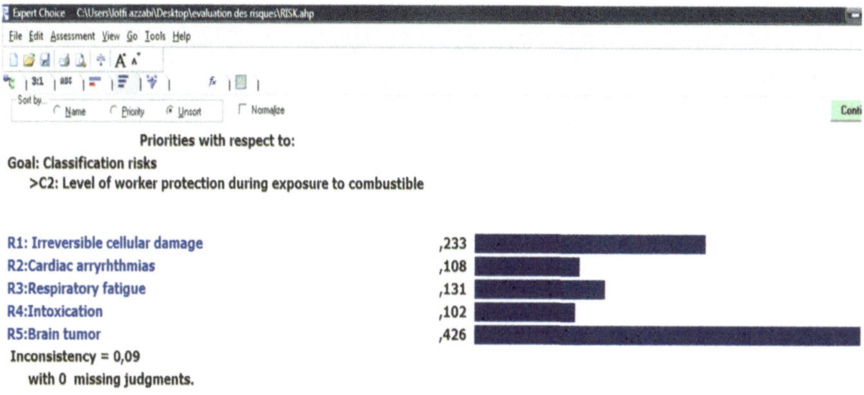

Fig. 5.13 Vector corresponding to the criterion C2 " Level of worker protection during exposure to combustible"

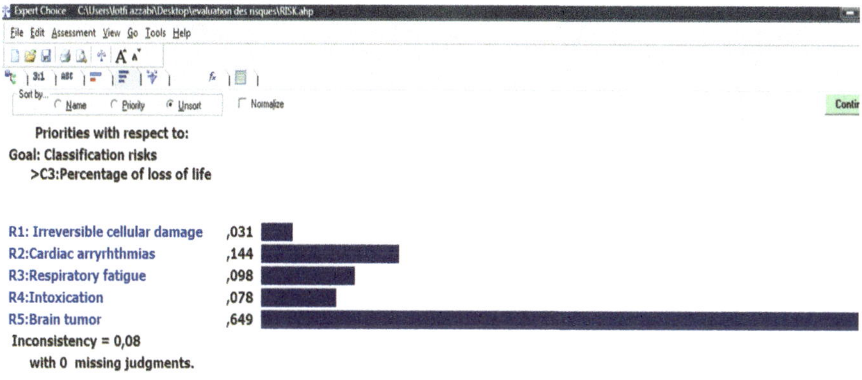

Fig. 5.14 Vector corresponding to the criterion C3 " Percentage of loss of life"

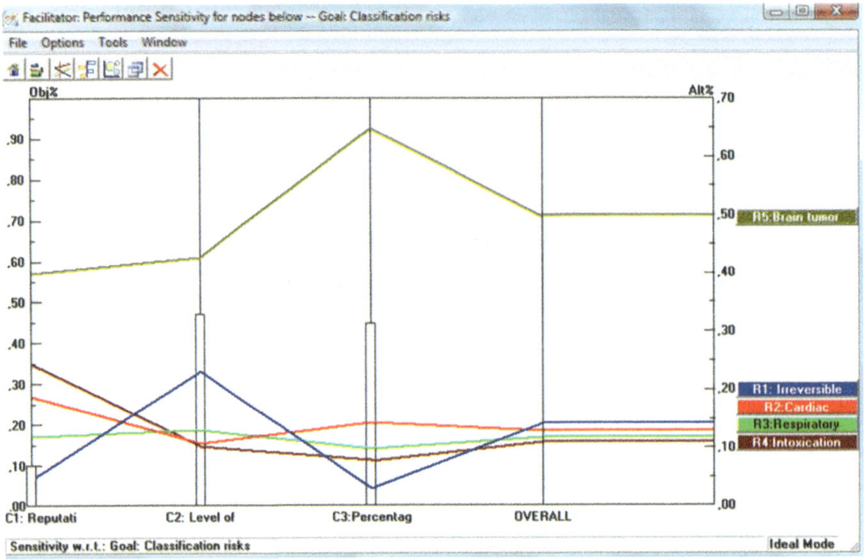

Fig. 5.15 Calculation of performance

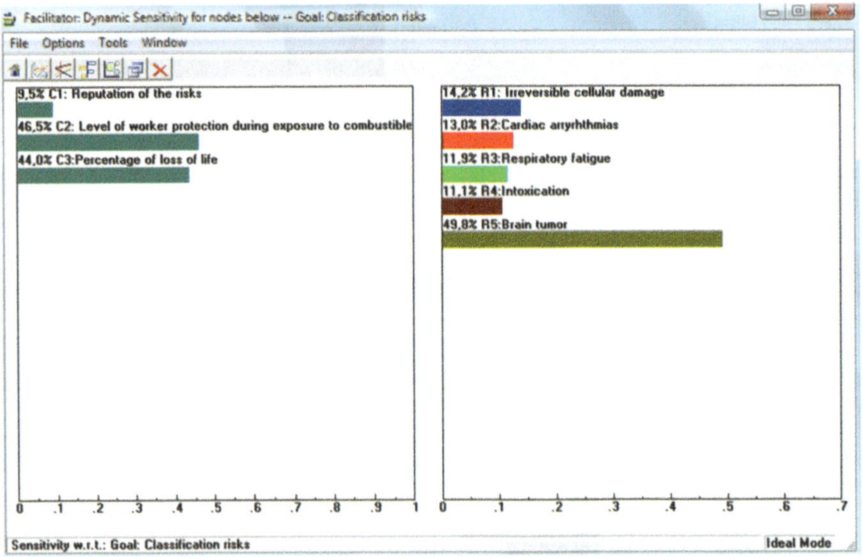

Fig. 5.16 Classification of risks

5.5.4 Record Your Finding

The protective measures taken on the basis of statistical taken during the 12 months of work in the plant processing gas are preventive measures to protect everyone in the vicinity of the fuel on a regular or occasional basis (Fig. 5.17).

Statistics risk of tumors that identifies the percentage of occurrence of this risk is very high is above the limits of acceptable risk levels set in the graph at 35%.

Then, the protective measures are (Table 5.7):

5.5.5 Review Your Assessment

The review assessment step consist to identify that the improvement proposed in the step record the finding is realized with successful, and it has to permit minimize the damage provide by the risk R5:Brain tumor.

In the graph follows, we consider as the Statistics percentage of risk R5: Brain tumor appears during the 12 months of the year to decline from the level limit of acceptable risk, just for the months of July and October there are still improvements to be redone (Fig. 5.18).

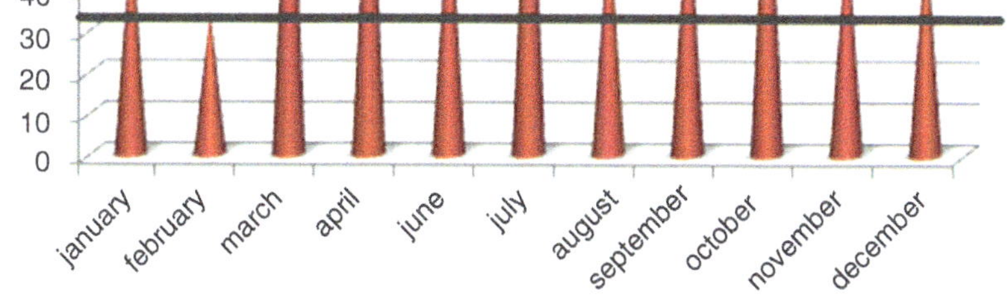

Fig. 5.17 Statistics of risks brain tumor

Table 5.7 Record the finding

Protective measures	Descriptions
A closed system	A closed system is a system allowing the maximum containment of substances used. Thus any contact between workers and these products can be avoided. The system can be defined as closed when all process operations in this respect containment: Transfer/transport of products, production, purification, cleaning and maintenance, sampling, analysis, treatment/disposal, storage ... Concretely, this can result in particular by a mechanical process, transform, or automation of certain tasks: Transfer of goods by mechanical or pneumatic sampling mechanized washing tanks without opening ... Be especially watchful for everything concerning the maintenance of such systems
Capture at source	Capture at source is a measure which is to channel the flow of pollutant emitted to a ventilation and disposal avoiding its release into the atmosphere of the workplace. This aspiration should be to the closest point of emission, in order to maximize system efficiency and minimize the flow required. It must be done using the natural movement of pollutants with air velocities sufficient and well distributed, draft-free, and parasite with an air intake compensation
The crib	The crib is set up physical barriers (walls, sides, cowl ...) that prevent the pollutant in question to spread in the atmosphere. It can be a total enclosure (glove box, fumc ...), occasionally with a possible opening for an operation inside the enclosure. It may also be a partial enclosure (single-wall ...) limiting and authorizing the issuance of air velocities lower catchment. The crib is always coupled with a measure of abstraction, it increases efficiency

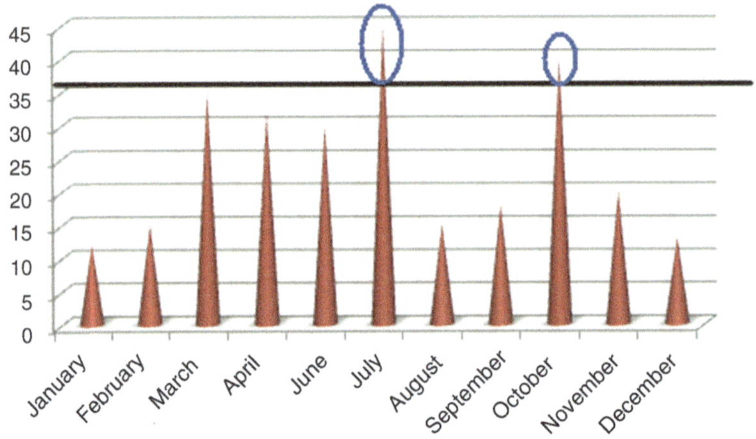

Fig. 5.18 Statistics of risks brain tumor after review

5.6 Conclusion

The study of the human risk in industry of treatment Gas, having capacities of reconfigure, or else specimens of dependences, comes for the tools utilization and methods very sophisticated in order to write down to the risks maximum and human errors.

The risk assessment must be "suitable and sufficient" to the proportion of risk, this should include making the premises inherently safe so that the routine and non-routine activities can be adequately assessed and controlled, e.g. employees working off site, mobile workers, home workers, and non-routine operations such as unplanned maintenance of equipment. Then, the challenge of risk assessment lies in those cases where complete information on these factors is unavailable. A conservative approach is generally advisable when insufficient information forces subjective judgement. Universal precautions are always advisable.

The advantage of the PROMETHEE and AHP method is modeling of the problem decision by hierarchic structure and the utilization of a semantic ladder to express preferences of the settler.

References

Ayadi, D., Azzabi, L., Kobi, A., Robledo, C., & Chabchoub, H. (2008). Classification of human risks with the method analytical hierarchy process. In Quality and dependability the 11th international conference, September.

Bernhard, K., Catharina, G., & Marc, F. (2007). State/event fault trees—A safety analysis model for software-controlled systems. *International journal of Reliability Engineering and system safety, 92*(11), 1521–1537.

Brans, J. P., Vincke, P. H., & Mareschal, B. (1986). How to select and how to rank projects: The Promethee method. *European Journal of Operational Research, 24*, 228–238.

Colson, G. (2000). The OR's prize winner and the software ARGOS: How a multijudge and multicriteria ranking GDSS helps a jury to attribute a scientific award. *Computers & Operations Research, 27*, 741–755.

Fadier, E., & Ciccotelli, J. (1999). How to integrate safety in design: Methods and models. *Journal of Human Factors and Ergonomics in manufacturing, 9*(4), 367–380.

Fadier, E., Garza, C., & Didelot, A. (2003). Safe design and human activity: Construction of a theoretical framework from an analysis of a printing sector. *Safety Science, 41*, 759–789.

Goetsch, D. L. (1997). *Implementing Total safety management*. New York: Simon & Schuster.

Gregoriades, A., & Sutcliffe, A. (2008). Workload prediction for improved design and reliability of complex systems. *Reliability Engineering and System Safety, 93*, 530–549.

Hale, A., Kirwan, B., & Kjellén, U. (2003). Safe by design: Where are we now. *Safety Science, 45*(2007), 305–327.

Health & Safety Executive. (1999). *HSG48, reducing error and influencing behaviour. HSE* Books.

Health and Safety Executive. (2006). Five steps to risk assessment. *INDG*163(rev2) 06/06.

Huacan, F., & Menglan, D. (2014). Safety System Engineering for Offshore Oil Offshore Operation Facilities Equipment and Procedures, Pages e183-e347.

Hughes, G., & Weichel, M. K. (2004). Whose fault is it anyway? A practical illustration of human factors in process safety. *Journal of Hazardous Materials, 115*, 127–132.

Huylenbroeck, G. V. (1995). "The conflict analysis method" bridging the gap between ELI CTRE, PROMETHEE and ORESTE. *European Journal of Operational Research, 82*, 490–502.
Ingram, S. (2003). Risk assessment. In *Fact sheet- Umatilla chemical agent disposal facility*. State of Oregon: Department of environmental Quality.
Larsen, R. F., & Buede, D. M. (2002). Theoretical framework the continuous early validation (CEAVA) method. *Journal Systems Engineering, 5*(3), 223–241.
Leena N, Paula S. (2004). Usability evaluation of complex system: a literature review. STUK-YTO-TR.
Marc, M.J. (1999). Multicriterion decision aid: methods and applications. In *CORS-SCRO, Annual conference, Windsor, Ontario*.
Neathey, F., Sinclair, A., Rick, J., Ballard William Hunt, J., & Denvir, A (2006). An evaluation of the five steps to risk assessment. In *Institute for Employment Studies Mantell Building Falmer Brighton BN1 9RF, HSE BOOKS*.
Neboit, M. (2003). A support to prevention integration since design phase: The concepts of limit conditions and limit activities tolerated by use. *Safety Science, 41*, 95–109.
Nijstad, B. A., Stroebe, W., & Lodewijk, H. F. (2003). Production blocking and idea generation: Does blocking interfere with cognitive processes. *Journal of Experimental Social Psychology, 39*, 531–548.
Osborn, A.F. (1963). *Applied imagination: Principles and procedures of creative problem solving* (3rd Revised ed.). New York: Charles Scribner's Sons.
Papadopoulos, Y. Maruhn, M. (2001). Model-based automated synthesis of fault trees from Matlab-Simulink models. In *DSN'01, Int'l Conf. on Dependable Systems and Networks* (pp. 77–82). Götenborg.
Pate-Cornell, M. E. (1998). Uncertainties in risk analysis: Six levels of treatment. *Reliability Engineering and System Safety, 54*(2-3), 95–111.
Haimes, Y. Y. (1996). *Risk Modeling, Assessment, and Management*. New York: Wiley.
Pierre, B. J., & Bertrant, M. (1986). How to select and how to rank projects: The PROMETHEE method. *European Journal of Operational Research, 44*, 1–10.
Rasmussen, J. (1999). The concept of human error: Is it useful for the design of safe systems. *Safety science monitor*, special edition.
Roy, B. (2000). Multicriteria aid for decisions. *A French–English decision aiding glossary*. Newsletter of the European Working Group Series 3, vol. 1.
Saaty, T. L. (1989). *Group decision making and the AHP in the analytic hierarchy process: Application and studies*. Berlin: Springer-Verlag.
Saaty, T. L. (2001). *The analytic network process: Decision making with dependence and feedback*. Pittsburgh, PA: RWS Publications.
Saaty, T. L. (2002). *Fundamentals of decision making with the analytic hierarchy process*. Pittsburgh, PA: RWS Publications. edition, revised.
Saaty, T. L. (2007). Time dependent decision-making; dynamic priorities in the AHP/ANP: Generalizing from points to functions and from real to complex variables. *Mathematical and Computer Modelling, 46*, 860–891.
Sylvain, D., Elie, F., Olivier, C., & Jean-Marc, P. (2003). Complexité des interactions entre un modèle de sécurité et un modèle d'organisation. *Qualita, -Swaminathan*.
Villemeur, A. (1992). *Methods and techniques, reliability, availability, maintainability and safety assessment*. New York: Wiley.

The manufacturer's authorised representative in the EU is Springer Nature Customer Service Centre GmbH, Europaplatz 3, 69115 Heidelberg, Germany. If you have any concerns regarding our products, please contact ProductSafety@springernature.com

Printed and bound by CPI Group (UK) Ltd, Croydon, CR0 4YY
30/09/2026
02225710-0002